Oceanic Fronts in Coastal Processes

Proceedings of a Workshop Held at the Marine Sciences Research Center, May 25-27, 1977

Edited by Malcolm J. Bowman and Wayne E. Esaias
and Coauthored by the Participants

Springer-Verlag
Berlin Heidelberg New York 1978

Malcolm J. Bowman
Wayne E. Esaias
Marine Sciences Research Center
State University of New York
Stony Brook, New York 11794/USA

The workshop was co-sponsored by:
United States Coast Guard, Environmental Protection Agency.
Office of Naval Research, New York Sea Grant Institute,
United States Department of Energy,
Marine Sciences Research Center, The Stony Brook Foundation.

ISBN 3-540-08823-7 Springer-Verlag Berlin Heidelberg New York
ISBN 0-387-08823-7 Springer-Verlag New York Heidelberg Berlin

Library of Congress Cataloging in Publication Data. Main entry under title: Oceanic fronts in coastal processes. Bibliography: p. Includes index. 1. Water masses--Congresses. 2. Coast changes--Congresses. I. Bowman, Malcolm J. II. Esaias, Wayne E. III. New York (State). State University at Stony Brook. Marine Sciences Research Center. IV. United States. Coast Guard. GC297.032 551.4'7 78-8917

Printing and binding: Beltz Offsetdruck, 6944 Hemsbach/Bergstr.
2132/3130-543210

To Michitaka Uda

SUMMARY AND RECOMMENDATIONS

Introduction

In this report we discuss the physical and biological properties of coastal fronts, including estuarine fronts. We assess the present state of scientific knowledge, what are the significant environmental implications, what are the most important areas upon which to focus future research, and what resources will be needed to attain those goals.

The Importance of Fronts

The study of oceanic fronts is an emerging science. Fronts are important in ocean dynamics since they are regions where exchange between different water masses are intense. Large scale fronts have important effects on the weather and climate. An understanding of their causes and effects is necessary in forecasting the ocean climate.

Fronts have important implications to underwater acoustic propagation and hence naval underseas efforts to describe, model and exploit mesoscale phenomenon to increase the effectiveness of fleet operations.

Oceanic fronts have significant biological consequences. They are areas of high productivity for all the food chain from phytoplankton through fish and marine mammals. They are therefore regions that have always been exploited by fishermen. Fisheries management strategies must take into account the distribution and productivity of coastal fronts. A clear understanding of the role of biological fronts is needed in the construction of biological models of ocean productivity.

The ability of surface fronts to concentrate ocean pollutants thousands of times has important and hitherto neglected implications for the uptake into the food chain of hazardous substances such as heavy metals, PCB's, etc. The location of persistent nearshore frontal zones must be considered in the design of waste water discharge outfalls including those for municipal, industrial, radioactive and power plant effluents into the marine environment if we are to avoid the excess concentrations and contamination of the shoreline.

Frontal circulation affects the dispersion of oil spills and ocean dumped contaminants. Thus, clearly, the study of coastal fronts must be incorporated into environmental monitoring and sampling design if we are to effectively determine how pollutants are transported, concentrated and incorporated into the marine food chain.

Present State of Knowledge

Our present lack of detailed understanding of the characteristics of coastal fronts is due to the notorious difficulty of making effective physical and biological field measurements, the avoidance of these difficulties by scientists in favor of studying more tractable problems, and the organizational and logistical obstacles in conducting extensive field programs. Also a factor is the lack of appreciation by program officers in governmental agencies charged with environmental management and exploitation of the important roles oceanic fronts play, plus the fragmented and overlapping interests of these agencies.

At present we possess sufficient knowledge to describe the gross features of frontal circulation, and to develop simple analytical models to describe horizontal and vertical motions. Frontal generation, meandering, persistence and dissipation are essential characteristics of fronts; our knowledge of these mechanisms is still limited.

The response of phytoplankton communities to favorable growing conditions in coastal fronts has received much attention, and we are beginning to understand and quantify these processes. We are still hampered in attacking problems at higher trophic levels by a lack of sampling techniques and realistic biological models to adequately determine species interactions and response rates. We do know that they are important.

The ubiquitous but ephemeral nature of coastal fronts invariably complicates the prediction of oil spill trajectories by mathematical models. We still have not been able to develop a useful, universal model for use in coastal and estuarine waterways which takes the collecting and concentrating influence of fronts into account.

We still are unable to confidently predict the subsurface motions of water near frontal zones from surface features as sensed by remote imagery.

Recommendations for Research

As part of our recommendations for future research we identified fourteen areas that deserve high priority. The following list is not necessarily in order of importance; further details can be found in section 2.4.

1. The development of conceptual and dynamical models to aid in setting up hypotheses, experimental design and the reduction of data.

2. The improvement of our measurements of the circulation near and along fronts.

3. The development of enough complementary techniques for the simultaneous examination of properties near and along fronts to gain a complete and common comprehension of both the physics and biology.

4. The identification and quantification of air-sea, internal and bottom exchange processes, the formation of new water masses, and mass balances important to the frontal ecosystem.

5. The development of techniques to infer the subsurface fields from surface or remotely sensed information.

6. The elucidation of the time evolution, persistence and dissipation of frontal zones.

7. The understanding of the interactions of frontal zones with various types of waves, tides and meteorological phenomena.

8. The determination of the essential dynamics of the formation of meanders and eddies associated with fronts.

9. The extension of present ecosystem modeling studies on the physical-biological coupling to higher trophic levels including the fish and benthos.

10. The search for new ways of improving our techniques for making biological rate measurements, such as productivity, grazing, nutrient uptake and diffusion.

11. The better quantification of the essential variability of frontal zones and the implication to the ecosystem.

12. The placing of higher priority to faster and better data processing to enable effective real time changes to be made in sampling strategies.

13. The investigation of the human health hazard in the uptake of toxic metals and other pollutants in frontal zones by fish.

14. The determination of the horizontal and vertical transport of ocean contaminants such as spilled oil, toxic and radioactive wastes, sewage, etc., in frontal zones.

Conclusion

Each of these research areas is sufficiently complex that a substantial increase in scientific attention and hence funding levels will be needed if significant progress is to be made. There needs to be a national commitment, and the subsequent organizational framework to follow, from both the scientific community and by governmental decision makers, to address these problem areas and to significantly increase our comprehension of the roles of oceanic fronts in coastal processes.

PREFACE

On May 25, 1977 a small invited group of coastal oceanographers assembled at the Marine Sciences Research Center at Stony Brook for three days of intensive discussions in a cloistered setting. The purpose of this workshop was to "assess the state of the art, to ascertain priorities for future research and to formulate the theoretical, instrumental, experimental and logistical tools needed to attain those goals in the study of coastal oceanic* fronts."

Although the existence of oceanic fronts has been known for a long time, ocean frontology is experiencing rapid acceleration in the emergence of new concepts and methodology. The science is developing from the descriptive phase and many unsolved problems lie in the understanding and quantification of frontal dynamics. In turn, challenging questions need to be addressed on the controlling influence of the physics of fronts on the chemistry, biology, acoustics, and suspended particulate aggregations in these zones.

Coastal fronts are very efficient at concentrating buoyant and suspended particulate matter including toxic wastes; heavy metal concentrations in polluted coastal frontal zones have been measured to be as high as one to ten thousand times background. These zones are also regions of high biological productivity, and consequently frequented by both commercial and sports fishermen. The inference is that fronts may represent sites where the uptake into the food chain of heavy metals, PCB's and other potentially harmful contaminants may be orders of magnitude higher than one might expect from averaged oceanic concentrations. Frontal zones also affect the dispersion of spilled oil, and an understanding of frontal physics is crucial to the development of oil dispersion models and to contingency planning for clean up operations.

For all the above reasons, we felt that a carefully focused and programmed workshop would represent a timely opportunity to critically assess our present understanding of coastal fronts, and provide some guidelines for future directions, both to the scientific community, and those program officers in governmental agencies concerned with basic research and marine environmental protection and management.

During the workshop special attention was given to frontal physics and the scale dependent coupling between these physical driving mechanisms and the biological responses at the lowest trophic levels. This, in part, reflected the scientific interests of the participants, but it was also, as mentioned above, due to the fact that frontal zones often represent important sites for food chain dynamics studies.

The summary of the workshop deliberations is followed by contributions co-authored by conferees. A deliberate review of the literature on coastal fronts has not been made. Instead "snapshots" of several categories of coastal frontal research projects have been provided. Obviously, there are omissions , but it is hoped that, at least in those subjects discussed, these chapters form a reasonably concise summary of the present state of the science of coastal fronts.

On behalf of all the participants we wish to express our appreciation to the various sponsors for making the workshop and these proceedings possible. We welcome any and all comments. We have taken the attitude that this is a progress report that will be improved and updated at some future time.

---

*"Oceanic" in this report is taken to include both open ocean as well as estuarine fronts.

Malcolm J. Bowman
Wayne E. Esaias

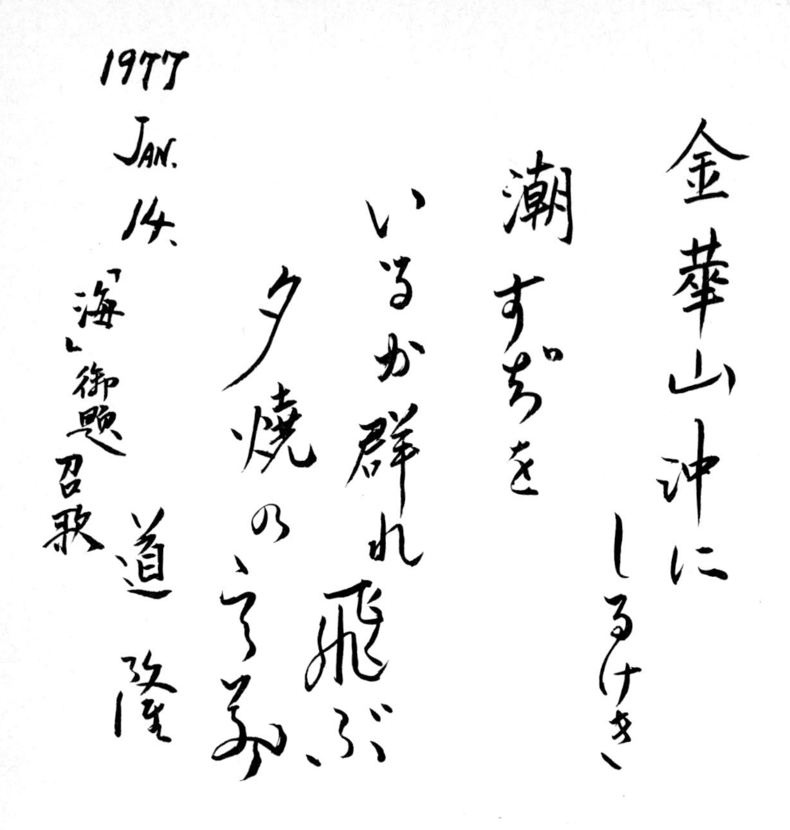

Michitaka Uda.

Translation: Porpoise school flying in the evening red sky along oceanic front in the offing of Kinkazazan!

(Invited poem presented at the traditional New Year's celebration held at the invitation of Emperor Hirohito, Imperial Palace, Tokyo. This ancient ceremony, with poetic contributions from leading citizens, can be traced back 1000 years.)

CONTENTS

Part I: Introduction and Proceedings

Photograph of advection fog arising at the Celtic Sea front in July 1975. Cold, vertically mixed water to the north of the front (background) is causing condensation of the air mass whose dew point has a value between the sea temperature on either side of the front. See Chapter 5 for further details (J. H. Simpson).

1. INTRODUCTION AND HISTORICAL PERSPECTIVE

MALCOLM J. BOWMAN

An oceanic surface front is the sea surface manifestation of a sharp boundary zone between adjacent water masses of dissimilar properties. The study of the physical, chemical, biological and optical properties of frontal zones makes up the discipline of frontology, the science of fronts.

Frontogenesis is the phenomenon of frontal generation and frontolysis, that of frontal dissipation.

Fronts form where there are horizontal variations in energy generation and dissipation processes (including kinetic, potential and thermohaline).

Important physical driving forces are those associated with air-sea transfers, including planetary and local wind stress, and seasonal and planetary vertical transports of heat (heating and cooling of sea surface) and water (evaporation and precipitation).

Other processes to be counted include riverine inputs of fresh water, confluences and shears of tidal and surface geostrophic flows, turbulent stirring due to the topography and roughness of the sea bottom, stirring due to internal wave and internal tide shear and nonlinear instabilities, and centrifugal effects due to curvature in the flow.

The dynamics of larger fronts are significantly influenced by the rotation of the earth, whereas small scale fronts are likely to be dominated by nonlinear inertial and frictional effects alone (although some dispute this).

Frontal boundary zones are regions of intensified motion and have a number of dynamically important characteristics, which will be discussed in later chapters. They also are regions rich in biological productivity and are thus favorite fishing grounds.

The scientific study of the cause and effects of oceanic fronts goes back to the middle of the 19th century. The American oceanographer, M. F. Maury, in 1858, described a front as a "wonderful phenomenon on the sea". G. F. Neumeyer, in 1875, supposed fronts to be "the collision and the struggle of two currents".

But the first documented account of fishing strategies near a front may refer back to the spring of A.D. 34. St. John records[1]: "Simon Peter said, "I am going out fishing". "We will go with you," said the others. So they started and got into the boat. But they caught nothing.

Morning came and there stood Jesus on the beach, but the disciples did not know it was Jesus. He called out to them, "Shoot the net to starboard and you will make a catch". They did so and found that they could not haul the net aboard, there were so many fish in it".

Frontal zones typically maintain a strong convergence of surface currents; in addition fronts limit the distribution of fish. It is fascinating to speculate that the disciples' boat was drawn into a surface convergence in the sea of Galilee and they were fishing in the biologically impoverished water mass to the port side. By casting their nets to starboard, they were able to draw in fish from the productive side of the front. Fronts (often called thermal bars) do form in large lakes during vernal warming and it could be scientifically and historically interesting to investigate this further.

The Japanese oceanographer, Michitaka Uda (1938,1959), to whom this document is dedicated, conducted pioneering descriptive and dynamical studies on oceanic fronts in the coastal seas around Japan.

In addition to classifying fronts both in dynamical and geomorphological terms (i.e., locations where different

frontal zones are likely to form) Uda provided a wealth of descriptive material, much gathered from the experience of Japanese fishermen.
Some examples:

At the boundary between water masses one often finds a visible line of demarcation, with peculiar ripples or waves which is a convergence (current or tide rip) or a divergence (oily slick) known as "siome".

Flotsam accumulates along the siome convergence line. Often this includes detritus such as dust, foam, timber and the whole food chain from phytoplankton up through zooplankton, molluscs, fish, birds, insects, dolphins, whales and finally man (in the forms of fishermen fisherpersons? and cadavers). Thus the siome becomes a productive fishing area.

The sea state in frontal zones can become quite violent, especially if the current shear across the front is large. Steep pyramidal standing waves can form when wave trains cross into a region of impeding current, or where a strong current flows opposite to a strong wind. In stormy weather a very dangerous sea state can occur with confused steep white capped waves reaching above deck level. In Japanese parlance this mature state is known as "sionami".

There is a fascinating account in Thor Heyerdahl's book on the Kon-Tiki expedition[2] on what was by all indications (although they did not realize it at the time) a crossing of a mid-oceanic front.

"There was not supposed to be land of any sort in the 4,300 sea miles that separated the South Sea Islands from Peru. We were therefore greatly surprised when we approached 100° west and discovered that a reef was marked on the Pacific chart right ahead of us on the course we were following. It was marked as a small circle, and, as the chart had been issued the same year, we looked up the reference in *Sailing Directions for South America*. We read that "breakers were reported in 1906 and again in 1926 to exist about 600 miles southwestward of Galapagos Islands, in latitude 6° 42' S., longitude 99° 43' W. In 1927 a steamer passed one mile westward of this position but saw no indication of breakers, and in 1934 another passed one mile southward and saw no evidence of breakers. The motor vessel 'Cowrie', in 1935, obtained no bottom at 160 fathoms in this position."...

For two days and nights we drove the raft north-northwest. The seas ran high and became incalcuable as the trade wind began to fluctuate between southeast and east, but we were lifted up and down over all the waves that rushed against us. We had a constant lookout at the masthead, and when we rode over the ridges, the horizon widened considerably. The crests of the seas reached six feet above the level of the roof of the bamboo cabin, and, if two vigorous seas rushed together, they rose still higher in combat and flung up a hissing watery tower which might burst down in unexpected directions. . . .

Next day the seas were less confused, as the trade wind had decided that it would now blow for a time from due east. We relieved one another at the masthead, for now we might expect to reach the point we were making for late in the afternoon. We noticed more life than usual in the sea that day, Perhaps it was only because we kept a better lookout than usual.

During the forenoon we saw a big swordfish approaching the raft close to the surface. . . . When we were having a rather wet and salty midday meal, the carapace, head, and sprawling fins of a large sea turtle were lifted up by a hissing sea right in front of our noses. When that wave gave place to two others, the turtle was gone as suddenly as it had appeared. This time too we saw the gleaming whitish-green of dolphins' bellies tumbling about in the water below the armored reptile. The area was unusually rich in tiny flying fish an inch long, which sailed along in big shoals and often came on board. We also noted single skuas and were regularly visited by frigate birds, with forked tails like giant swallows, which cruised over the raft. Frigate birds are usually regarded as a sign that land is near, and the optimism on board increased. ...

From noon onward Erik was more and more diligent in climbing up on the kitchen box and standing blinking through the sextant. At 6:20 P.M. he reported our position as latitude 6° 42' south by longitude 99° 42' west. We were 1 sea mile due east of the reef on the chart. The bamboo yard was lowered and the sail rolled up on deck. The wind was due east and would take us slowly right to the place. When the sun went down swiftly into the sea, the full moon in turn shone out in all its brilliance and lit up the surface of the sea, which undulated in black and silver from horizon to horizon. Visibility from the masthead was good. We saw breaking seas everywhere in long rows, but no regular surf which would indicate a reef or shoal. No one would turn in; all stood looking out eagerly, and two or three men were aloft at once.

As we drifted in over the center of the marked area, we sounded all the time. All the lead sinkers we had on board were fastened to the end of a fifty-four-thread silk rope more than 500 fathoms long, and, even if the rope hung rather aslant on account of the raft's leeway, at any rate the lead hung at a depth of some 400 fathoms. There was no bottom east of the place, or in the middle of it, or west of it. We took one last look over the surface of the sea, and when we had assured ourselves that we could safely call the area surveyed and free from shallows of any kind, we set sail and laid the oar over in its usual place, so that wind and sea were again on our port quarter."

The frontal zone is inherently unstable, and meanders of various scales (ranging from $\sim$100m in estuaries and shallow seas to $\sim$500km in the Gulf Stream) form, grow and may spin off eddies into either or both water masses. Japanese fishermen record the existence of vortices or whirlpools several meters in diameter in continental shelf zones, which are very effective in rolling up fishing nets into a ball and sucking them down like an inverted tornado.

Uda also records some fascinating details on the acoustic properties of fronts. The irregular wave rippling near relatively quiescent fronts has a boiling noise ("butu-butu, chabu-chabu, saa-saa") similar to the sound of water in a rushing stream, attributed to the combination of breaking bubbles and the striking sound of the splash during violent rippling.

Sionami is accompanied by a recognizable noise ("zaa-zaa") which can heighten to a roaring noise ("goo-goo") in the presence of breaking pyramidal waves, and may be heard for several miles.

Early investigators soon became aware of the difficulty of making current measurements in frontal zones and the need for multiple ship surveys to gather synoptic data. Much use was made of fishermen's reports on the location and characteristics of fronts during the fishing season in the western Pacific Ocean and the Japan Sea (one of the recommendations in this report is to use fishermen to document the location and characteristics of shallow sea fronts).

Uda was quick to realize that siome is seldom a simple, single demarcation or boundary, but usually is compound in nature with imbedded, intermittent, multiple features such as bands of strong and weak gradients, striae and various scales of motion. He demonstrated the usefulness of contouring vorticity and convergence in the surface layer as a mapping tool in identifying regions where frontogenesis and lateral shear (and hence the potential for generation of eddies are likely. He also predicted that benthic (bottom) convergence should be a driving mechanism for the formation of sand banks and drifts.

In recent review, Roden (1976) stressed the physics of oceanic fronts and the prediction of frontogenesis. He pointed out that recent advances are, in large part, due to the development of radiative transfer theory and satellite imagery. Successfully used satellite methods include sensed sea surface temperature contrasts by very high resolution infrared radiometry (spatial and amplitude resolution ~0.5km and ~0.5C, respectively), and specular optics which rely on the detection of sea state discontinuities across fronts. False color imagery which detects differences in suspended particulate matter has proven to be useful for quantifying both suspended sediment and chlorophyll *a* concentrations in coastal waters, when ground truth measurements are available for calibration. The combined use of satellites, aircraft and ships have been found very useful in the study of coastal currents, especially in estuaries where fronts are ubiquitous features.

Following the proceedings of the workshop, there are nine contributions co-authored by the conferees. While the organization by subject matter may not be perfect, it reflects the expertises of the authors and represented the most efficient way of rapidly publishing this report.

We have not attempted to provide exhaustive documentation of all relevant references. However, each contributed chapter is followed by a short list of important references to the subjects discussed. Mooers, et al. (1976) is a comprehensive bibliography on oceanic fronts, stressing physical aspects. This will be updated in 1978. At the workshop, R. Fournier was inspired to create a biological bibliography on fronts. That will be a more difficult task since the advertences to fronts in the biological literature are much more diffuse. Contributions to and requests for copies of these bibliographies should be addressed directly to the authors (addresses are given in Appendix B).

---

[1] The New English Bible, p. 141, Oxford University Press, Cambridge University Press, 1970 pp.

[2] Thor Heyerdahl, 1973. Kon-Tiki. Across the Pacific by Raft. Ballantine Books, pp 144-146.

References

Maul, G. A. and H. R. Gordon. 1975. The use of earth resources technology satellite (LANDSAT-1) in optical oceanography. Remote Sensing Environment, 4, 95-128.

Mooers, C. N. K., T. B. Curtin, D. P. Wang, and J. F. Price. 1976. Bibliography for oceanic fronts and related topics: Special project from RSMAS, University of Miami, Florida.

Uda, M. 1938. Researches on "siome" or current rip in the seas and oceans. Geophys. Mag., 11, 307-372.

Uda, M. 1959. Seminar 2: Water mass boundaries - "siome" frontal theory in oceanography. Fisheries Res. Board Canada Manuscript Report Section 51 (unpublished).

Rodgers, G. K. 1965. The thermal bar in the Laurentian Great Lakes. Publ. 13. Great Lakes Research Division, University of Michigan, 358-363.

Roden, G. I. 1976. On the structure and prediction of oceanic fronts. Naval Research Reviews, 29 (3), 18-35.

2. PROCEEDINGS OF THE WORKSHOP

2.1 *What Is An Oceanic Front?*

A universal, concise definition of an oceanic front is difficult to set down. Rather, the nature of fronts are best perceived through a community of characteristics.

Historically, fronts have been characterized by fishermen as narrow bands or streaks on the sea surface with peculiar wave or rippling characteristics and commonly associated with filaments of foam and floating matter. The Japanese term for front "Siome" means "current or tide rip." Others have described a front as the collision line of two ocean currents, often accompanied by intensified horizontal and vertical motions and mixing.

Frontal zones, representing the boundary between horizontally juxtaposed water masses of dissimilar properties, are a fundamental feature of geophysical turbulence and play important roles in ocean dynamics. They are regions of convergence and relatively strong vertical motions. Frontal circulation can represent a mechanism for cascading energy from large to small scales. They are delineated by singularities in horizontal gradient and/or higher derivates of temperature, salinity, density, velocity, sea state, chlorophyll, etc.; i.e. their location is defined as the position of a singularity (maximum horizontal gradient and/or higher derivatives) of one or more of the above characteristics.

Fronts occur on all spatial scales, from fractions of a meter to global extent. The largest fronts, formed in regions of convergence of planetary scale winds, have important effects on weather and climate.

Fronts are found in the surface layers, at mid depths, and near the ocean bottom (benthic fronts). Fronts can conveniently be classified into six categories:

1) Fronts of planetary scale, usually associated with the convergence of surface Ekman transports, and found away from major oceanic boundaries (e.g., within the Sargasso Sea, Southern Ocean).

2) Fronts representing the edge of major western boundary currents. These fronts are associated with the intrusion of warm, salty water of tropical origin into higher latitudes (e.g., Gulf Stream, Kuroshio).

3) Shelf break fronts formed at the boundary of shelf and slope waters, such as are found in the Middle Atlantic Bight. Circulation along the front may or may not be baroclinic depending on whether the temperature and salinity fronts coincide, and whether their contributions to the density field are compensating or reinforcing.

4) Upwelling fronts, essentially the surface manifestation of an inclined pycnocline, commonly formed during a coastal upwelling, i.e. as a result of an offshore surface Ekman transport associated with alongshore wind stresses (e.g., along the west coast of the U. S., Peru, Northwest and Southwest Africa).

5) Plume fronts at the boundaries of riverine plumes discharging into coastal waters (e.g., Amazon, Columbia, Hudson, Connecticut, etc.).

6) Shallow sea fronts, formed in continental seas and estuaries, and around islands, banks, capes, shoals. These are commonly located in boundary regions between shallow wind and tidally mixed nearshore waters and stratified, deeper, offshore waters (e.g., Celtic and Irish Seas, approaches to English Channel, Long Island Sound).

All of these frontal systems share common properties of persistence, ranging from hours to months in spite of diffusion of properties across strong horizontal gradients, and surface convergence with associated strong vertical convection, usually at least an order of magnitude greater than open ocean vertical convection.

The component of flow parallel to fronts frequently has intense horizontal shear in a direction normal to the front. Such shears may be in a geostrophic equilibrium for the larger fronts, but flows near shallow water, small scale fronts are expected to be more strongly influenced by local acceleration, bottom stress, and by interfacial friction than Coriolis force.

In this report, we purposely restrict our attention to the latter four categories. This is partly due to the scientific interests of the participants as coastal oceanographers, and partly due to our concern with man's effects on the coastal environment and the role fronts play in the distribution of various trophic levels of marine organisms and pollutants.

2.2 *Why Are Fronts Important?*

Fronts are important in ocean dynamics since they are regions where vertical advection and the exchange of momentum and other properties are locally intense. Fronts form between adjacent water masses as well as between the ocean interior and the surface and bottom boundary layers. Large scale surface fronts are important in air/sea interaction effects on the weather and climate. In this regard, oceanic forecasting will probably never be satisfactory until we can resolve and parameterize the dynamic features of these fronts, which at the present time usually lie in the sub grid scale of numerical circulation models.

Fronts are environmentally important in that one or two sided surface convergences are very effective in collecting and concentrating floating detritus, and other particulate matter. Heavy metal concentrations in convergence zones have been measured to be 3 or 4 orders of magnitude above background in polluted coastal waters. Oil slicks commonly line up along surface convergences, and affect the dispersion of spilled oil, both horizontally and vertically.

A knowledge of the characteristics of persistent local fronts may be important to the design and positioning of sewer and power plant heated effluent outfalls, positioning of oil rigs, and the dumping of ocean contaminants including radioactive wastes.

The location of persistent frontal zones may be important to agencies charged with search and rescue, since small boats, swimmers and cadavers can easily be drawn into convergence zones. Formation of sea fog at coastal fronts can, under certain conditions, become a hazard to navigation.

The design of fishing strategies for maximum yields involves a detailed knowledge of the locations of oceanic fronts, since these have always been found to be regions of high biological productivity. Phytoplankton blooms which often form in restabilizing water masses near frontal zones can develop, under favorable conditions, into toxic red tides. Further, survival strategies of fish on the continental shelves presumably include a response to the temporal and spatial scales of coastal fronts. It is commonly known that fish orient themselves toward fronts, regions rich in buoyant particulate organic matter, but often, unfortunately, also rich in pollutants.

A clear understanding of the role of biological fronts is clearly needed in both diagnostic and prognostic biological productivity models. Biologists clearly need to be thinking about ecosystem models incorporating frontal zones as important features while actively interacting with those physical oceanographers who are researching frontal dynamics.

Underwater acousticians have documented, observationally and theoretically, that open ocean fronts do serve as acoustic lenses, leading to propagation loss anomalies of orders of magnitude both plus and minus. The question naturally arises of whether or not acoustic methods, using fixed or moving sources and receivers, could be used effectively as supplementary devices for monitoring and probing oceanic fronts, even in the coastal ocean. It is well known that layers of biological and geological particulate matter, and temperature and salinity microstructure are prominent in frontal zones. There also exists some documentation of acoustic scattering from these layers. Thus, acoustic scattering information, especially from research vessels, may provide a useful reconnaisance and research tool in frontal research, including the coastal ocean. Since marine mammals and other acoustically active creatures congregate at coastal fronts, frontal zones can be expected to be anomalously intense in ambient noise. The fact that fronts can be self-insonified might be used to advantage in their study.

Finally, the study of coastal fronts must be incorporated in environmental monitoring and sampling design programs, if we are to effectively determine how pollutants are transported, concentrated and incorporated into the food chain in the marine environment and if biased and aliased sampling is to be avoided.

2.3 *What Are The Major Problems In Expanding Our Understanding Of Coastal Fronts?*

The study of oceanic fronts is still in its infancy. There are many reasons for this, both historical and scientific.

First, the field measurements necessary for the development of dynamic models are notoriously difficult to make. These include measurements of cross and along front velocities, both in the horizontal and vertical, shear stresses and entrainment rates at the frontal interface, meandering, persistence and resolution of maximum gradient regions. Further weaknesses in our observational data lie in the lack of repeated sampling (repeated sections, anchor stations, etc.) for frontal evolution information, and lack of along-front synoptic information.

Second, there has been a lack of priority in the physical community given to the study of fronts, and a tendency to concentrate on problems which appear to be more tractable.

Third, we can never rely on a single set of experiments since there are multiple scales and temporal evolution involved. To improve our understanding of the dynamics, chemistry and biology of fronts, there is a need to develop suites of instrumentation. Experimental platforms will include multiple ships, towed systems, aircraft, satellites, moored systems, and submersibles.

The design and conduct of imaginative field experiments, and the concurrent evolution of conceptual frameworks and diagnostic and prognostic models, will require substantial support. The development and deployment of expensive arrays of new instrumentation will be required. The expense of such instrumentation, which may include the cost of extensive arrays of current meters, microstructure profilers, acoustic remote sensors, towed fluorescence/nutrient samplers, is far above the funding levels of the usual grants made to individuals, particularly in universities. There is no way of avoiding the necessity of developing complex and costly research programs if we are to significantly advance the state of art in the science of coastal fronts.

Fourth, there is an urgent need to educate the scientific community, as well as elected officials and the program managers of the funding agencies on the importance of coastal fronts, both in oceanography and environmental management.

The study of fronts is one of the outstanding areas in oceanography where interaction between physicists and biologists is important. This may call for a readjustment of priorities in funding agencies to give more emphasis to interdisciplinary research. For example, research in oceanography at the National

Science Foundation is funded through the traditional disciplines. It is more difficult to obtain funding for interdisciplinary projects.

There are too many governmental agencies with vested, overlapping and fragmented interests in the coastal ocean to make a coherent research program on coastal fronts possible. (In this regard there will be an opportunity for dialogue on these matters at the Chapman Conference on Oceanic Fronts.) It would be useful if agencies could begin to collaborate on frontal research and a myriad of other problems in the coastal ocean. On the other hand university researchers and academic institutions are often less than enthusiastic about making long term commitments to large scale, multi-year, multi-institution endeavors.

2.4 *Where Do We Go From Here? Some Future Goals.*

The conferees were able to list a host of challenging scientific problems that should be addressed. There was a consensus, for example, that a large outer shelf frontal study may be five or ten years away. Meanwhile, as a practical matter, we have to be opportunistic by designing experiments with the use of ships and aircraft of opportunity, satellite imagery, and the use of casual observers (e.g. fishermen) to build up our descriptive knowledge of fronts. It was also concluded that we should emphasize the continuing study of small scale fronts driven by inertial and frictional effects since much of the instrumentation already exists, and the problems are more tractable.

Since large scale fronts have various dynamic scales, it is necessary to inbed a series of small scale experiments into the large scale experiment. We must iterate towards the most ambitious experiments. Also it is much more difficult to place instrumentation arrays (e.g. current meters) in deep water as compared to the continental shelf and estuaries.

The following is a list of goals (they are not all independent) established for future coastal frontal research.

1) The development of conceptual and dynamical diagnostic (to digest data) and prognostic (to aid in setting up hypotheses and experimental design) models of frontal zones. Attempts to use isentropic analysis and potential vorticity conserving dynamics in rationalizing data fields should be useful (see chapter on frontal dynamics). Calculation of momentum, heat, and salt fluxes, and their divergences should be given emphasis.

2) To improve our measurement of the circulation near and along fronts, including within maximum velocity gradient regions. Lagrangian and Eulerian methods need to be used.

3) To develop enough different tools to simultaneously examine fronts to gain a complete and common comprehension of both the physics and biology. Underway sampling systems may be one of the more promising approaches to synoptic studies of fronts.

4) To identify and quantify exchange processes: air-sea, internal, and bottom which may include *bona fide* turbulence measurements, and the formation of new water masses. The determination of the various mass balances are especially important for ecosystem and other non-physical effect studies.

5) To find techniques to infer the subsurface fields from surface or remotely sensed information (e.g., temperature, color, directional wave field, surface height).

6) To determine the time evolution of the formation (frontogenesis), persistence and dissolution (frontolysis) of frontal systems.

7) To determine the interaction of frontal regimes with various wave classes, e.g. surface gravity waves, inertial-internal waves, topographic Rossby waves,

and planetary Rossby waves.

8) To determine the essential dynamics of cyclogenesis and anticyclogenesis in frontal zones; i.e., to determine how detached cyclonic and anticyclonic eddies are generated from fronts.

9) To extend present ecosystem modeling studies on the physical/biological coupling to higher trophic levels including the benthos and fish, since most work to date has concentrated on phytoplankton dynamics in frontal zones.

10) To find ways of improving our techniques for making biological rate measurements, such as productivity, grazing stress, nutrient uptake, diffusion, etc., in frontal zones. These rate measurements are likely to be more important at the higher trophic levels than at lower levels. Development of an *in situ* micro autoanalyser might be given priority; it should be deployable together with current meters in order to make circulation flux and turbulent flux measurements.

11) To characterize the essential variability of frontal zones. For example, the scale and variance of horizontal and vertical water velocities have profound effects on phytoplankton accumulation, growth, and mortality. Especially important (and confounding) are those scales of variability comparable to the doubling time of phytoplankters (~ 1 day). For example, transient upwelling effects, vertical mixing and subsequent restabilization of the water column during and after storm events, or, in the case of shallow sea fronts, during the restabilization phase of the spring-neap tidal stirring cycle, can each lead to spectacular plankton blooms with subsequent decay. Such bursting/relaxation effects stress the need for physical-biological *time dependent* models.

12) To give priority to more rapid data processing. For example, the sophistication of telemetry systems is rapidly increasing, and now makes possible real time data analysis. Experiments need to be carefully planned with sufficient lead time or rapid data analysis between cruises to allow for informed changes in sampling strategies.

13) To investigate the human health hazard of the uptake of toxic metals and other pollutants in frontal zones by commercial species of fish.

14) To determine the horizontal and vertical transport of oceanic contaminants such as spilled oil by coastal fronts, and to develop techniques of oil spill trajectory forecasting, which include the effects of fronts.

2.5 *A Suggested Scenario For A Comprehensive Frontal Study*

The following outline is a logical approach to an ambitious, five year theoretical and field study of coastal fronts. It is intended as a guide to how such a program could be organized to fulfill some of the goals outlined in the preceeding section.

It is essentially multidisciplinary in nature, and the active participation of physicists, biologists, chemists and geologists, would be necessary throughout. Physicists would have to lead the way in developing the physical structure.

To ensure rapid progress a number of individuals would need to be dedicated to managing the project to maintain effective coordination of its various aspects. These managers would have to give high priority throughout to model development and experimental design. Considerable effort would have to be devoted to planning, especially to phasing of research and long lead time developments, and to communications inside and outside of the project. These models (incorporating physical, chemical, biological and geological aspects) would hopefully lead to the identification of a few key parameters to be measured for model verification.

For example, the measurement of physical entities such as frontal convergence, interfacial friction and entrainment, wind driven spin up or spin down rates, distances between the surface convergence zone and the frontal line itself, etc., could alleviate many detailed measurements.

These key measurements could be used to calibrate frontal circulation models, which then would be used to predict the velocity field in the frontal zone.

Timetable: Note: year long "spacers" might be necessary to digest recent results before proceeding to the next step.

Year 1: Conduct seasonal, high resolution, three dimensional hydrographic and biological synoptic surveys to obtain scale and variability information. The use of multiple ships would be desirable for synopticity, and extensive use would be made of satellite and aircraft remote imagery.

Year 2: In addition to repeating the seasonal studies, synoptic scale current meter arrays would be developed to generate long time series at strategic locations. Pilot studies of fine scale hydrography (e.g., inversions, interleaving intrusions, turbulence dissipation rates), drogue and biological rate measurements would be accomplished in these strategic locations.

Year 3: To the above seasonal studies, mesoscale (~100 km) and boundary layer (both surface and bottom) current meter arrays would be developed, to investigate atmospheric forcing and bathymetric effects. Fine scale measurements of interleaving, turbulence, diffusion, productivity, species selective grazing experiments, nutrient recycling, etc. could be made.

Year 4: Repeat year 3 studies, but with more emphasis on seasonal transition periods when frontogenesis/frontolysis are likely to be intermittent processes.

Year 5: Conduct a year long study, with emphasis given to repeating key measurements for model initializations, calibrations, and verifications.

2.6 *Some General Comments And Caveats On Sampling Strategies*

During the workshop, many comments were made, based on the collective experience of the participants regarding sampling strategies. The following discussion is certainly not exhaustive, but should be useful in developing future programs.

Sampling strategies for oceanic fronts need to consider two easily overlooked factors: one, fronts are multiple scale phenomena, which implies different dynamical balances for various scales, and two, a comprehensive set of physical fields need to be sampled in order to distinguish the frontal dynamics.

2.6.1. The Problem of Resolution and Synopticity

The recognition of the existence of fronts in many regimes was slow to develop due to the lack of spatial resolution. After the dawn of recognition, the maximum horizontal gradients were still underestimated due to a need for further horizontal resolution (~ 0.1 to 1 km). Similarly, the interleaving of layers was missed until the vertical resolution was substantially increased (~ 0.5 to 5 m) over classical methods. As the temporal resolution has increased in the proximity of fronts, strong internal tidal, near-inertial motion, and high frequency internal gravity wave activity have been noted. (So far, apparently, the only turbulence measurements made have been a small scale, direct observation dye dispersion study.[1])

The easily neglected complement to resolution for the smallest scales involved is domain (span or duration) for the largest scales. In the cross-front direction, the span is typically of the order of a few baroclinic radii of deformation, $R_{bc} = (g'D)^{\frac{1}{2}}f^{-1}$, where $g' \sim 0.5$ to $4\ cm\ sec^{-2}$ is the reduced gravity, $D \sim 2 \times 10^3$ to 2×10^4 cm is the far field thickness of the density layer, and $f \sim 10^{-4}$ sec^{-1} (mid-latitude value) is the Coriolis parameter. Thus, $R_{bc} \sim 3$ to 30 km. In the alongfront direction, the span is determined by the predominate meander wavelength, which in turn is determined by bathymetric scales, atmospheric forcing scales, wavelength of coastally-trapped waves, or baroclinic instability scales. Though the prevalent dynamics and scales in this direction are not well known, a likely range of values is 30 to 300 km. The vertical span should probably be a few mixed layer thicknesses, say 60 to 600 m. The duration should be at least one or a few evolutionary cycles, typically 10 to 100 days.

It is difficult to acquire synoptic measurements in frontal research. Eulerian instrument arrays give gross information in frontal zones, since frontal meandering (which may be of the order of minutes [estuarine fronts] to a week [shallow sea and shelf break fronts]) will advect the frontal interface across and out of the fixed array. Near surface current records may be contaminated with surface wave motion. Long term averaging of velocity records will blur the details of frontal circulation. (Two day running averages are typically needed to resolve shelf break frontal meandering.)

Remotely sensed Lagrangian drogues have been successfully used in river plume frontal studies, referenced to a coordinate system moving with the front. Such studies could be extended to the shelf-break front.

At first glance, the sampling requirements may seem frightening, but, due to the singular nature of the front *per se*, non-uniform spacing in the cross-front and vertical directions, samples clustered at strategic locations in the alongfront direction, and continuous measurements in time can be used. The use of remote acoustic sensing, vertically profiling current meters and rapidly profiling STD measurements go a long way towards the goal of synopticity.

Obtaining the complementary biological measurements is much more difficult. The goal of rapid biological rate measurements remains elusive. The use of variable trim towed photometers, nephelometers, fluorometers and pumping systems for the continuous measurement of distributions of light, chlorophyll "a" and nutrients have proven their usefulness in standing stock assessment. The development of a real time, continuous zooplankter counter would be significant.

2.6.2. The Use of Nested Arrays

Further progress on frontal dynamics will dictate the development of instrumentation to simultaneously determine both micro and mesoscale dynamics. One method of achieving this goal would be the deployment of nested or clustered "antennas" or arrays, where the finer scales of motion are measured in the center, with the outer components having sufficient span and duration to resolve the largest scales. Such arrays may be designed to be nonuniform in the horizontal, vertical, and time dimensions, and need not necessarily be Eulerian in nature. For example, aircraft drops of instrument packages into strategic locations during the experiment might prove useful. (The conceptual design of such a nested array suitable for the study of estuarine dynamics has recently been published by Kinsman *et al.*[2])

2.6.3. Agency Procurement of Instrumentation

Since the instrument and logistical requirements for significant advances in the science of coastal fronts are very expensive, and well beyond the resources of any one institute, the workshop recommended the establishment of instrument banks, perhaps by the National Science Foundation. NSF presently owns and operates several oceanographic research vessels which are block funded for NSF grantees. Perhaps NSF should also acquire a suite of instruments such as profiling and fixed current meters, hyperbolic and satellite navigation systems, towed systems, *in situ* fluorometers, automatic nutrient analysers, acoustic remote sensors, STD recorders, etc. Such instruments are certainly needed for frontal research and many other areas.

Major problems associated with such a bank were perceived to include the chronic high mortality rate of oceanographic instruments, maintenance and logistical problems. Of course, if the bank is professionally managed, those existing problems are solved!

---

[1]Pingree, R. D., G. R. Forster and G. K. Morrison. 1974. "Turbulent convergent tidal fronts." J. Mar. Biol. Assn. U. K., 54:469-479.

[2]Kinsman, B., J. R. Schubel, M. J. Bowman, H. H. Carter, A. Okubo, D. W. Pritchard and R. E. Wilson. 1977. TRANSPORT PROCESSES IN ESTUARIES: Recommendations for Research. Marine Sciences Research Center, Special Report 6.

Part II: Contributions by Participants

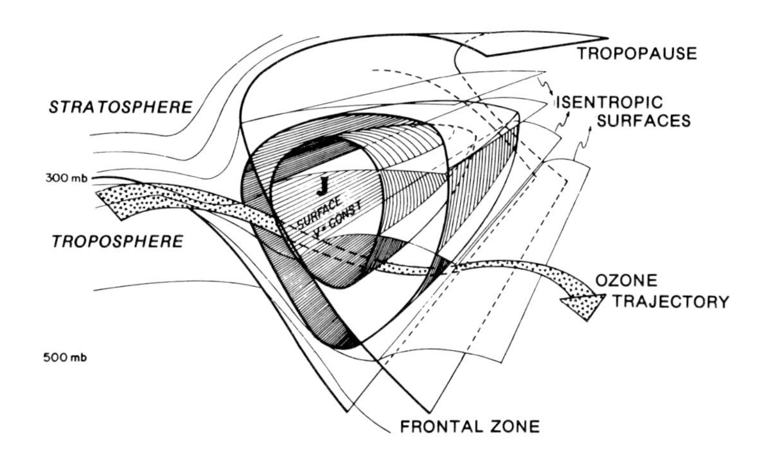

A three dimensional schematic of the atmospheric jet stream; its frontal zone; and a representative trajectory for ozone transfer from the stratosphere, across the tropopause via the frontal zone, and into the troposphere. This inspiration from meteorology is indicative of one of the transfer routes in frontal zones which oceanographers might expect to find important. (From the brochure cover for Conference on Air Quality Meteorology and Atmospheric Ozone; July 31-August 6, 1977; University of Colorado at Boulder.)

3. FRONTAL DYNAMICS AND FRONTOGENESIS

CHRISTOPHER N. K. MOOERS

Frontal Dynamics

Introduction. The dynamics of oceanic fronts are not yet fully established, but it is clear that there are several types of fronts with somewhat different dynamics. A density front embedded in a geostrophic flow is considered here as an important case. This case may correspond to density fronts and geostrophic flows found on the outer continental shelf and over the continental slope, as well as in the open ocean. The objective is to illustrate what may influence the cross-stream circulation in such a case. It is possible to formulate time-dependent models for the generation of fronts (frontogenesis) which are time-dependent and non-dissipative, but steady-state front models and models for annihilation of fronts (frontolysis) require dissipation. (The latter requires both time-dependence and dissipation.) Thus, to keep the discussion general, both dissipation and time dependence are incorporated in the formulation.

Analysis. Without providing the scale analysis here, a candidate, reduced system of equations for a rotating (f-plane), stratified fluid is presented for the interior (i.e., away from surface and bottom boundary layers), with the front assumed to be oriented along the y-axis:

$$- fv = - \frac{1}{\rho_o} \partial_x p \tag{1}$$

$$\partial_t v + u\partial_x v + w\partial_z v + fu = - \frac{1}{\rho_o} \partial_y p + \partial_z (N_v \partial_z v) \tag{2}$$

$$0 = - \frac{1}{\rho_o} \partial_z p + b \tag{3}$$

$$\partial_x u + \partial_z w = 0 \tag{4}$$

$$\partial_t b + u\partial_x b + w\partial_z b = \partial_z (K_v \partial_z b) \tag{5}$$

where b is the buoyancy (equivalent to density) $\partial_i = \partial/\partial i$, and all other variables are quite standard, including f, the Coriolis parameter, N_v, the vertical eddy viscosity, and K_v, the vertical eddy diffusivity. These equations are derived with the assumption that the alongfront scale is much greater than the cross-front scale and that the cross-front velocity is much smaller than the alongfront velocity.

From (1) and (3), the alongfront flow is in geostrophic balance and the regime is in hydrostatic balance. All the physical complexity is in (2) and (5), especially the former which has a role for the local and advective acceleration ("nonlinear effects") of alongfront flow, alongfront pressure gradient, and vertical mixing of alongfront momentum in driving cross-front flow. The nonlinear terms are expected to be important in frontal zones where the local Rossby ($\partial_x v$ f) and Richardson $\partial_z b/(\partial_z v)^2 = f^2/\partial_z b(\frac{\partial z}{\partial x}|b)^2$ numbers are of order one.

Several things can be done with this system of equations. For example, if, as is commonly the case, v and b are known as functions of (x, z, t), $\partial_y p$ is given, and the frictional terms can be estimated, (2) and (5) can be solved diagnostically (algebraically) for u and w; then, the continuity equation (4) can be applied as a constraint, i.e., as a consistency condition for the cross-front flow. As a stronger example, since (1) and (3) imply that the thermal wind relation ($f\partial_z v = \partial_x b$) holds for all time, then (2) and (4) can be cross-differentiated to eliminate the time dependence:

$$\partial_z[u\, f\,(f + \partial_x v) + wf\partial_z v] - \partial_x\,[u\partial_x b + w\partial_z b] = \partial_z\,[\partial_z(N_v f\partial_z v) - \partial_x(K_v\partial_z b)], \tag{6}$$

where it has been assumed that the alongfront pressure gradient is barotropic (i.e., uniform with depth). It is convenient to introduce the stream funciton, ψ, and a pressure-like function Q such that

$$u = -\partial_z\psi \text{ and } w = \partial_x\psi;\; Q = \frac{p(x,z)}{\rho_0} + \frac{f^2x^2}{2};\; f\,(f + \partial_x v) = \partial^2_{xx}Q,$$

$f\partial_z v = \partial^2_{xz}\,Q = \partial_x b$, and $\partial_z b = \partial^2_{zz}\,Q$; then (6) reduces to

$$\partial^2_{xx}Q\;\partial^2_{zz}\psi - 2\,\partial^2_{xz}Q\;\partial^2_{xz}\psi + \partial^2_{zz}Q\;\partial^2_{xx}\psi = \partial_z\left[\partial_z(N_v\partial^2_{xz}Q) - \partial_x\,(K_v\partial^2_{zz}Q)\right] \tag{7}$$

Notes. (i) The terms with first order partial derivatives of ψ vanish due to the symmetry of the factors involving partial derivatives of Q.

(ii) $N_v = K_v$ = constant is a sufficient condition for the righthand side to vanish.

(iii) This is an elliptic problem for ψ if $\Delta = \partial^2_{zz}Q\;\partial^2_{xx}Q - (\partial^2_{xz}Q)^2 > 0$ (the usual case), where Δ is essentially the absolute potential vorticity ($\Delta < 0$ is the necessary condition for symmetric baroclinic instability), which can be solved diagnostically once Q (v and b), the mixing terms, and boundary conditions for ψ are specified.

An attractive feature of (7) is that the cross-front circulation (u and w) appears as a secondary flow induced by inertial (nonlinear) and internal frictional effects involving only the specified buoyancy and alongfront flow fields. Additional driving is obtained from the boundary conditions, as discussed below. Since (7) is second order in z, only two vertical boundary conditions can be applied. Again, this provides a diagnostic model for ψ.

For prognostic purposes, (1), (2), (3) and (5) can be combined to form:

$$\frac{D}{Dt}\,\partial_x Q = -\frac{f}{\rho_o}\,\partial_x p + \partial_z\,(N_v\partial^2_{xz}Q) \tag{8}$$

$$\frac{D}{Dt}\,\partial_z Q = \partial_z\,(K_v\partial^2_{zz}Q), \tag{9}$$

where $\frac{D}{Dt}(\cdot) = \partial_t(\cdot) + u\partial_x(\cdot) + w\partial_z(\cdot)$ and (4) still constrains u and w.

From (8) and (9), it is straightforward to show that

$$\frac{D}{Dt}(\Delta) = \partial^2_{zz}Q\;\partial^2_{xz}\,(N_v\partial^2_{xz}Q) + \partial^2_{xx}Q\;\partial^2_{zz}\,(K_v\partial^2_{zz}Q) - \partial^2_{xz}Q\left[\partial^2_{xz}\,(K_v\partial^2_{zz}Q) + \partial^2_{zz}\,(N_v\partial^2_{xz}Q)\right] \tag{10}$$

which is the absolute potential equation for this problem. If there is no diffusion, Δ is conserved along particle paths in the (x, z) plane. (Pedlosky, 1977, used this formulation [Δ = const.] to create a theory of coastal upwelling frontogenesis.)

Alternatively, if there is no time-dependence, this is a (nonlinear) elliptic problem for Q (u and w can be eliminated between [8] and [9]) as long as

$$N_V K_V \partial^2_{xx} Q \partial^2_{zz} Q - \frac{(N_V + K_V)^2}{4} (\partial^2_{xz} Q)^2 > 0,$$

which is assured if $N_V = K_V$ and $\Delta > 0$. Since (10) is fourth order in z, four vertical boundary conditions should be satisfied. Similarly, because it is second order in x, two horizontal boundary conditions should be satisfied.

The boundary conditions are derived with the aid of time-dependent Ekman layer equations:

$$\partial_t u_E - f v_E = \partial_z (N_V \partial_z u_E) \tag{11}$$

$$\partial_t \bar{v}_E + (f + \partial_x v_I) u_E = \partial_z (N_V \partial_z u_E) \tag{12}$$

$$\partial_x u_E + \partial_z w_E = 0 \tag{13}$$

$$0 = - \frac{1}{\rho_o} \partial_z p_E + b_E \tag{14}$$

$$\partial_t b_E + u_E \partial_x b_I = \partial_z (K_V \partial_z b_E) \tag{15}$$

where $u = u_I + u_E$, u_I, etc. are the interior solutions discussed earlier, and u_E, etc. are the Ekman layer solutions. The surface (z = 0) boundary conditions are that

$$\psi = \psi_I + \psi_E = 0, \quad N_V (\partial_z v_E + \partial_z v_I) = \frac{\tau_w^{(y)}}{\rho_o}, \quad N_V \partial_z u_E = \frac{\tau_w^{(x)}}{\rho_o},$$

$$K_V (\partial_z b_E + \partial_z b_I) = B_s, \text{ and } p = p_E + p_I = p_a.$$

Thus, (11) and (12) yield

$$\partial^2_{tt} u_E + f(f + \partial_x v_I) u_E = \partial_z \left[f(N_V \partial_z v_E) + \partial_t (N_V \partial_z u_E) \right]$$

and, upon vertical integration and appropriate substitution, then,

$$\frac{1}{f^2} \left[\partial^2_{tt}(\cdot) + f(f + \partial_x v_I) \right] \psi_I \Big|_{z=0} = \frac{\tau_w^{(y)}}{\rho_o f} - \frac{N_V \partial_z v_I}{f}\Big|_{z=0} + \frac{\partial_t \tau_w^{(x)}}{\rho_o f^2} \,. \tag{16}$$

The middle term on the right had side of (16) is the "Ekman choking" effect of fronts. Through the thermal wind relation, this term is non-vanishing if a surface density front exists; it serves to either reduce or enhance the effect of the alongfront wind stress on the cross-front surface Ekman transport. In the coastal upwelling case, the Ekman transport is reduced. Also, the horizontal shear of the interior alongfront flow in the upper layer alters the effective inertial frequency, $f(1 + \partial_x v_I/f)^{\frac{1}{2}}$, and, thus, the Ekman suction ($\psi_I|_{z=0}$).

Similarly, (14) and (15) yield, upon vertical integration and appropriate substitution,

$$-\frac{1}{\rho_o}\partial_t p_I|_{z=0} + \psi_I \partial_x b_I|_{z=0} = B_s - K_v \partial_z b_I|_{z=0} - \frac{1}{\rho_o}\partial_t p_a \,, \tag{17}$$

where the horizontal advection of buoyancy is seen to enter in the surface buoyancy balance through the second term.

Notes. (i) These boundary conditions can be expressed in terms of Q_I (and its partial derivatives) and ψ_I.

(ii) They are also nonlinear and have to be integrated in time.

(iii) ψ_I can be eliminated between (16) and (17).

(iv) Analogous equations arise from the bottom ($z = -H$) boundary conditions;

$$\frac{1}{f^2}\{\partial^2_{tt}(.) + f(f+\partial_x v_I)\}\ \psi_I|_{z=-H} = \frac{\tau_B^{(y)}}{\rho_o f} - \frac{N_v \partial_z v_I|_{z=-H}}{f} + \frac{\partial_t \tau_B^{(x)}}{\rho_o f^2} \tag{18}$$

and

$$-\frac{1}{\rho_o}\partial_t p_I|_{z=-H} + \psi_I \partial_x b_I|_{z=-H} = K_v \partial_z b_I|_{z=-H}, \tag{19}$$

where $(\tau_B^{(x)}, \tau_B^{(y)})$ is the bottom stress and the buoyancy flux is assumed to vanish at the bottom.

In solving (7) diagnostically for ψ, $Q(x, z, t)$ would be given by b and v, (16) and (18) would be integrated in time to obtain the upper and lower boundary conditions for ψ respectively, and an appropriate condition on ψ would be imposed on the horizontal boundaries of the domain.

Note. (17) and (19) could not be imposed; at best, they could be used as a consistency check on Q or to adjust the observed Q at the boundaries.

In solving (7), (8) and (9) prognos tically for Q and ψ, (16) to (19) would be integrated for one time step and $\psi_I|_{z=0}$ and $\psi_I|_{z=-h}$ eliminated. Then (8) and (9) would be integrated for one time step and the spatial problem for Q would be solved subject to the conditions on horizontal and vertical boundaries. Next, (7) would be solved for ψ subject to the conditions on horizontal and vertical boundaries, and so forth through subsequent time steps.

Discussion. From the above analysis it can be appreciated that knowledge of the eddy viscosity and diffusivity functions is a crucial part of this problem. Determination of reasonable functional laws for N_v and K_v is always difficult, but it is especially so in frontal regions where mixing of mass and momentum can be expected to be intense, localized, and intermittent. From the author's point of view, it is more physically realistic to neglect horizontal mixing of mass and momentum and to allow for horizontal variations of vertical mixing, which can be relatively intense in the frontal zone.

Various observations support the notion that vertical mixing is associated with "mean flow" shear instabilities in the frontal zone. Here "mean flow" shear means the slowly-varying (compared to f^{-1}) thermal wind plus low frequency inertial-internal wave shear.

Johnson (1977) has shown that $N_v = N_v(R_i)$, where R_i is the Richardson number based on the thermal wind, provides a vertical structure to the term $\partial_z(N_v \partial_z V)$

which can produce a "double-cell" circulation pattern in the coastal upwelling frontal zone. He succeeded with various functional forms but $N_V \propto \frac{1}{R_i}$ was as successful as any. He considered a station on mid-shelf (near the frontal zone) and one on the edge of or off the shelf in each of four cases and compared his results with observed crossfront velocities. He obtained a relatively large N_V at the base of the pycnocline in the frontal zones. (He did not explicitly introduce a horizontal dependence nor did he consider nonlinear terms.) Due to the vertical variation of $N_V \partial_z V \propto (\partial_z V)^3/N^2$, a "double-cell" (or internal Ekman layer) flow was produced where generally observed. This mechanism of frictional cross-stream flow is much like that of the Thompson (1974) two-layered, two-sided entrainment model and the Garvine (1974) integral model for fronts which imposes downward turbulent entrainment at the interface in the frontal zone.

In a numerical model for the annual cycle of shallow sea fronts, James (1977) used a similar vertical eddy viscosity and diffusivity law (inverse dependence on R_i) but with a direct dependence on the wind stress (for simulation of surface stirring) plus tidal transport (for simulation of bottom stirring), which decay away from the surface and bottom, respectively. Again, the results were promising. (It is noted *en passant* that Endoh [1977a,b] has provided results from similar numerical experiments for both density and thermohaline fronts in coastal regions.)

These pioneering attempts deserve to be followed up with more rigorous theories, numerical experiments, and field experiments specifically designed to address turbulent exchange in frontal zones.

Frontogenesis

Hoskins and Bretherton (1972) noted that there are at least eight mechanisms important in changing temperature gradients and forming atmospheric fronts (a process called frontogenesis):

(1) horizontal deformation field,
(2) horizontal shearing motion,
(3) vertical deformation field,
(4) differential vertical motion,
(5) surface friction,
(6) turbulence and mixing,
(7) latent heat release, and
(8) radiation.

In the ocean, the first six mechanisms can be important, but the last two are generally negligible. For oceanic frontogenesis, horizontally nonuniform buoyancy fluxes (heating and cooling, precipitation and evaporation, river run-off, ice melt, ice brine, etc.) can be added to the list.

Oceanic frontogenesis, as the ocean itself, can be driven from its (frictional) boundaries, the generally most dynamically active of which is the surface boundary even in coastal regions. For the sake of illustration, frontogenesis in the upper layer is chosen for focus.

Frontogenetical equations. Some mechanisms for producing fronts (frontogenesis) in the upper (mixed) layer of the ocean are examined. Thus, the density, ρ_s, of the upper layer is considered to be only a function of the horizontal coordinates (x, y) and time (t). The upper layer extends from the sea surface ($z = \eta[x,y,t]$, the time and space averaged value of η is zero) to the base of the mixed layer ($z = -h[x,y,t]$) which is not a material surface. The fluid is assumed to be in hydrostatic balance; hence, in the upper layer the pressure is $p(x,y,z,t) = \rho_s g(\eta - z) + p_a$, where p_a is the atmospheric pressure.

For simplicity, it is assumed that the horizontal flows in the upper layer are not strongly sheared, at least initially which allows the neglect of advective accelerations. For convenience, lateral boundaries and mixing zones are considered to be sufficiently distant that the effects of horizontal diffusion can be neglected.

The (linearized) horizontal momentum equations for the upper layer are then

$$\rho_s(\partial_t u - fv) = -\partial_x p + \partial_z \tau^{(x)} \quad (1)$$

and

$$\rho_s(\partial_t v - fu) = -\partial_y p + \partial_z \tau^{(y)} \quad (2)$$

where $(\tau^{(x)}, \tau^{(y)})$ is the turbulent vertical flux of horizontal momemtum. The equation of continuity and the heat and salt equations are combined with a linearized equation of state to form

$$\partial_t \rho_s + \partial_x(\rho_s u) + \partial_y(\rho_s v) + \partial_z(\rho_s w) = \partial_z m^{(z)} \quad (3)$$

where $m^{(z)}$ is the turbulent vertical flux of density. Equations (1) to (3) are then integrated over the mixed layer to obtain

$$\rho_s(\partial_t u - f\tilde{v}) = -\partial_x \tilde{p} + (\tau_W^{(x)} - \tau_I^{(x)})/h \quad (4)$$

$$\rho_s(\partial_t \tilde{v} + f\tilde{u}) = -\partial_y \tilde{p} + (\tau_W^{(y)} - \tau_I^{(y)})/h \quad (5)$$

$$\partial_t \rho_s + \partial_x(\rho_s \tilde{u}) + \partial_y(\rho_s \tilde{v}) = M_s - \frac{\Delta\rho}{h} W_e, \quad (6)$$

where $\tilde{u} = \frac{1}{h}\int_{-h}^{\eta} u\,dz$, etc., $\vec{\tau}_I$ is the interfacial stress, M_s is the surface mass flux (heating and cooling; evaporation and precipitation), W_e is the turbulent entrainment velocity at the base of the mixed layer, and $\Delta\rho$ is the density difference between the mixed layer and the interior. By analogy to the entrainment flux of density, the interfacial stress can be parameterized as

$$\vec{\tau}_I = \rho_s \frac{\Delta\vec{v}}{h}$$

Note. In (4) and (5),

$$\nabla_H \tilde{p} = \nabla_H \rho_s g \frac{\eta}{2} + \rho_s g \nabla_H \eta + \nabla_H \rho_a,$$

which are baroclinic, baratropic, and atmospheric pressure gradient terms, respectively.

With the introduction of the differential operator

$$D(.) = \partial^2_{tt}(.) + f^2(.),$$

$\tilde{u}$ and $\tilde{v}$ can be separated between (4) and (5) and substituted into (6), which can then have the horizontal gradient operator, $\nabla_H(.)$, applied to it to yield:

$$\partial_t \nabla_N \rho_s = \nabla_H \{D^{-1}\left[\partial_t \nabla^2_H \tilde{p} - \partial_t(\nabla_H \cdot \vec{\tau}_W) - f(\nabla x \vec{\tau}_W)_v + \partial_t(\nabla_H \cdot \vec{\tau}_I) + f(\nabla x \vec{\tau}_I)_v\right] - \frac{\Delta\rho}{h} W_e + M_s\}. \quad (7)$$

The mixed layer depth, h, is also a prognostic variable. The turbulent closure assumption of hard-limiting shear stability can be used; i.e., let $R_i = (\Delta\rho/\rho_s \cdot gh)/|\Delta\vec{v}^2|$, then

$$\partial_t h \begin{cases} = 0, & R_i \geq R_{ic} \\ > 0, & R_i < R_{ic} \end{cases}$$

where R_{ic} is some 0(1) critical value.

The righthand side of Eq. (7) is the *frontogenetical function.* Thus, the surface density gradient will change if there is a horizontal gradient in the surface buoyancy flux, turbulent entrainment, or the D^{-1} time integral of the time rate of change of the horizontal Laplacian of pressure, wind or interfacial stress divergence or of the curl of the wind or interfacial stress. Whether frontogensis, frontolysis, or nothing occurs depend upon the sign of the sum of these effects.

References

Endoh, M. 1977. Double-celled circulation in coastal upwelling. J. Oceanographical Society of Japan, 33, 6-15.

Endoh, M. 1977b. Formation of thermohaline front by cooling of the sea surface and inflow of the fresh water. J. Oceanographical Society of Japan, 33, 30-37.

Garvine, R. W. 1974. Dynamics of small-scale oceanic fronts. J. Phys. Oceanogr., 4 (4) 557-569.

Hoskins, B. J. and F. P. Bretherton 1972 Atmospheric frontogenesis models: Mathematical Formulation and Solution. J. of Atmospheric Sciences, 29 (1), 11-37.

James, I. D. 1977. A model of the annual cycle of temperature in a frontal region of the Celtic Sea. Estuarine and Coastal Marine Science (in press).

Johnson, D. R. 1977. On double-cell circulation during coastal upwelling. (In preparation)

Pedlosky, J. 1977. A non-linear model of the onset of upwelling. (Preprint)

Suginohara, M. 1977. Upwelling and two-cell circulation. J. Oceanographical Society of Japan, 33, 115-130.

Thompson, J. D. 1974. The coastal upwelling cycle on a Beta-plane: Hydrodynamics and Thermodynamics. Ph.D. Dissertation, Florida State University.

4. ADVECTION - DIFFUSION IN THE PRESENCE OF SURFACE CONVERGENCE

AKIRA OKUBO

Introduction

Frontal zones are regions of convergence associated with relatively strong vertical motions. They also contain singularities in the horizontal velocity field such as lines of convergence, to which an infinite number of streamlines converge. A complicated velocity structure thus exists in the frontal zone.

Spatial heterogeneity in velocity or velocity gradients play a crucial role in oceanic diffusion (Bowden 1965). In particular, floatables such as oil and other pollutants tend to be trapped by convergences, and accordingly advection and diffusion of these particles in the presence of convergence is an important oceanographic problem with dynamical, biological and environmental implications (Klemas and Polis, 1977). This contribution is concerned with theoretical aspects only of quantifying some of these phenomena.

Horizontal Flow Patterns Near Surface Convergences

In the neighborhood of a surface convergence the horizontal velocity field (u,v) can be approximated linearly in horizontal coordinates (x,y) whose origin is taken at a point within the convergence zone. The velocity field is given by:

$$\begin{aligned} u &= 1/2\{(-c+\alpha)x + (h-\eta)y\} \\ v &= 1/2\ \{(h+\eta)x - (c+\alpha)y\} \end{aligned} \tag{1}$$

where c, α, h and η, are respectively the convergence rate (or negative divergence), stretching deformation rate, shearing deformation rate and vorticity (Okubo, 1970); viz.,

$$\text{convergence} \quad \frac{\partial u}{\partial x} + \frac{\partial v}{\partial y} \equiv -c$$

$$\text{stretching deformation} \quad \frac{\partial u}{\partial x} - \frac{\partial v}{\partial y} \equiv \alpha$$

$$\text{shearing deformation} \quad \frac{\partial v}{\partial x} + \frac{\partial u}{\partial y} \equiv h$$

$$\text{vorticity} \quad \frac{\partial v}{\partial x} - \frac{\partial u}{\partial y} \equiv \eta$$

The convergence rate is interpreted as the fractional rate of sea surface areal contraction in the horizontal plane. The stretching deformation rate implies change of shape by different rates of stretching along the coordinate axes. If α is positive, there is stretching along the x-axis and shrinking along the y-axis. The opposite holds for negative α. The shearing deformation rate is the difference between the two Cartesian components of stretching. Vorticity is a measure of the rate of rotation of a horizontal areal element.

The flow pattern in the vicinity of convergence can be classified by the roots of the characteristic equation of (1), i.e.

$$\begin{vmatrix} -c+\alpha-2\lambda & h-\eta \\ h+\eta & -c-\alpha-2\lambda \end{vmatrix} = 0 \tag{2}$$

or

$$\begin{aligned} \lambda_1 &= 1/2\{-c+(\alpha^2+h^2-\eta)^{1/2}\} \\ \lambda_2 &= 1/2\{-c(\alpha^2+h^2-\eta^2)^{1/2}\} \end{aligned} \tag{3}$$

Thus the behavior of trajectories of floatables placed in the flow field depends on the values of $\alpha^2+h^2-\eta^2$ and c.

If the roots of λ_1 and λ_2 are distinct, one has a solution for the trajectory of a fluid element initially located at (x_0,y_0) near the origin:

$$\begin{pmatrix} x \\ y \end{pmatrix} = \tfrac{1}{2}\begin{pmatrix} x_0 + (\alpha x_0+hy_0-\eta y_0)q & x_0 - (\alpha x_0+hy_0-\eta y_0)q \\ y_0 + (hx_0+\eta x_0-\alpha y_0)q & y_0 - (hx_0+\eta x_0-\alpha y_0)q \end{pmatrix} \begin{pmatrix} e^{\lambda_1 t} \\ e^{\lambda_2 t} \end{pmatrix} \tag{4}$$

where $q \equiv (\alpha^2+h^2-\eta^2)^{-1/2}$.

One may write a similar solution for multiple roots. By varying the initial fluid element, a family of trajectories is obtained, and, depending upon the nature of the roots λ_1 and λ_2, we have the following description of the trajectories in the vicinity of the origin as t increases.

(i) If λ_1 and λ_2 are real, and if

(a) $\lambda_2<\lambda_1<0$, then all trajectories approach the origin as a limit, forming a stable node there,

(b) $0<\lambda_2<\lambda_1$, the origin is an unstable nodal point, since the trajectories move away from it without limit,

(c) $\lambda_2<0<\lambda_1$, the trajectories other than those lying on the principal axes of the rate of strain quadric tend asymptotically to a set of generally oblique asymptotes, and hence, the motion is unstable; the origin forms a saddle point,

(d) $\lambda_2<0 = \lambda_1$, the trajectories approach a line described by $y = -(-c+\alpha)(h-\eta)^{-1}x$ if $h \neq \eta$ or $x = (c+\alpha)(h+\eta)^{-1}y$ if $h \neq -\eta$; the origin becomes no longer an isolated stable singularity, but every point of the line forms a stable singularity (a 'line of convergence'),

(e) $0 = \lambda_2<\lambda_1$, the trajectories move away from the line mentioned above, so that the origin as well as all other points of the line may be characterized as unstable singularities; the line represents a''line of divergence',

(f) $\lambda_1 = \lambda_2$, the trajectories may be straight lines, parabolas, or somewhat complicated logarithmic curves; the motion is seen to be stable if $\lambda_1 = \lambda_2<0$ and unstable if $\lambda_1 = \lambda_2 \geq 0$.

(ii) If λ_1 and λ_2 are complex conjugates, and if

(a) the real part is negative, then the trajectories are spirals that wind around and approach the origin as a limit; the origin is a stable spiral point,

(b) if the real part is positive, the spirals unwind and the trajectories move away from the origin without limit; the origin is an unstable spiral point,

(iii) If λ_1 and λ_2 are purely imaginary, the trajectories form ellipses with centers at the origin; the motion is thus a stable one about the origin, which may be called a vortex point.

Figure 1 summarizes the results stated above in diagrammatic form. The abscissa represents the relative importance between the stretching-shearing deformations and the vorticity, while the ordinate represents the convergence of the field. On the right side of the thick solid line, there are *unstable* saddle points of singularity at the origin ($x = y = 0$), while on the left side of it, there are *stable* nodal and spiral points of singularity at the origin. A portion of the thick line coincident with the negative axis of $\alpha^2+h^2-\eta^2$ represents *stable* vortex points of singularity at the origin. The remaining part of the thick line forming a half parabola represents the situation where a line convergence passing through the origin exists, and hence where the motion is *stable*.

Dispersion of Floatables in the Vicinity of Singularities

The sea surface is taken as a horizontal plane. Consider the horizontal dispersion of floatable particles in the neighborhood of singularities. Two distinct cases will be examined. One is the dispersion or rather deformation of particles in a linear velocity field in the absence of any random motions, and the other is the dispersion of particles in the presence of random motions superposed upon the velocity field.

(i) No random motions are present.

Let us place a group of floatable particles, initially arranged in a circle of radius r_0, near a singular point. The subsequent motion of each particle is

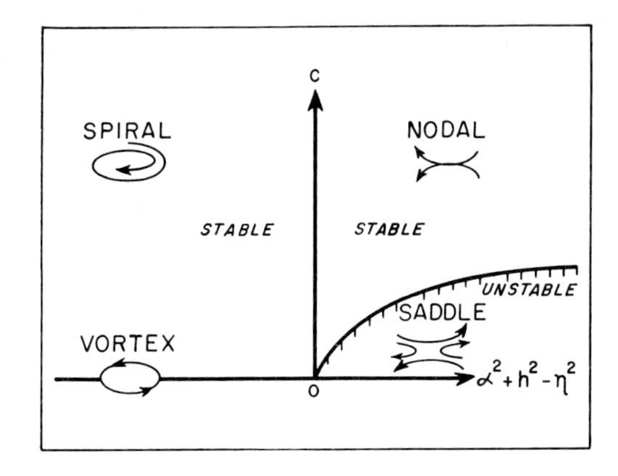

Fig. 1. Classification of singularities for the linear velocity field.

described by (4) subject to the constraint

$$(x_0-x_p)^2 + (y_0-y_p)^2 = r_0^2 \; ,$$

where x_p and y_p denote the coordinates of the center of the initial circle.

It can be shown that the initially circular pattern of particles will be deformed into an ellipse, the area of which is given by $\pi r_0^2 \exp\{-ct\}$. The area increases or decreases exponentially with time according to whether the velocity field has divergence or convergence. The motion of the center of mass of the particles can be brought into the same category as described previously (see Fig. 1).

Each particle approaches the origin asymptotically if it is a stable nodal or spiral point, and the area covered by these particles asymptotically shrinks into the point of singularity. On the other hand, when the origin is characterized as an unstable nodal point, each particle tends to move away from the origin, and the area increases without limit. Two different situations occur if the origin is a saddle point, i.e., unstable singularity. Although, in both cases, the floatable particles form asymptotically an extremely elongated ellipse, its major axis increases with time faster than its minor axis decreases when there is a divergence ($c<0$) and vice versa when there is a convergence ($c>0$). The net effect is an asymptotic increase in the area when $c<0$ and an asymptotic decrease in the area when $c>0$. As the area of floatables is shrinking, the group tends to move away from the converging point or line. The predominance of the deformation rates over vorticity and convergence rate is responsible for the behavior. If the origin is a vortex point ($c=0$), the particles turn around it, changing their overall shape periodically but maintaining their total areas the same.

(ii) Random motions are present.

We shall now add random motions to the linear mean-velocity field and investigate the horizontal dispersion of floatable particles from an instantaneous source. For simplicity, assume that the horizontal diffusive process due to the random motion can be described by a constant eddy diffusivity, D.

Let $n(x,y,t)$ denote the number of floatables per unit surface area, then n satisfies the following equation of advection-diffusion.

$$\frac{\partial n}{\partial t} + \frac{\partial un}{\partial x} + \frac{\partial vn}{\partial y} = D\left(\frac{\partial^2}{\partial x^2} + \frac{\partial^2}{\partial y^2}\right)n \; , \qquad (5)$$

where we assign to u and v the linear form specified by (5).

An analytical solution of (5) can be found for appropriate initial and boundary conditions (Okubo, Ebbesmeyer, Helseth and Robbins, 1976). Instead, I shall discuss the behavior of moments associated with the distribution of n. To this end, define (i,j) moments by

$$\overline{x^i y^i} \equiv \frac{1}{N} \iint_{-\infty}^{\infty} x^i y^i \, n \, dx \, dy \quad (i,j = 0,1,2,\ldots),$$

where N denotes the total number of particles. The first moments describe the position of the center of mass of particles, the second moments about the center of mass describe the variances of the particle displacements, and so forth. The equations for these moments can be derived directly from (5) (Okubo, 1970). If the mean variance grows without limit, we speak of the instability of variance, while if the mean variance approaches a certain limiting value at large times, we speak of the stability of variance. In the latter case, a group of particles attains a finite size as a limit.

The criteria for stability or instability of the variance turn out to be very much the same as those for trajectories in the linear velocity field described previously.

Floatable particles approach, as a group, a stable nodal or spiral point of singularity to attain a certain asymptotic group-size as a limit of time. In this case, the collecting power of the field convergence is so strong that the group size of particles cannot grow indefinitely notwithstanding the process of advection diffusion, i.e., the combined action of the velocity inhomogeneities and the diffusive process due to random motions. On the other hand, floatable particles asymptotically move away from an unstable nodal or spiral point of singularity and disperse without limit until the higher-order terms, of u and v become important and the linear theory becomes no longer valid. In the case of vortex points of singularity, a group of floatable particles will move around the singularity and, at the same time, the mean variance will grow linearly with time as a limit. In the presence of a line of convergence, floatables tend to line up on the convergence and ever disperse along it.

The dispersion of floating objects on some water surface, ocean or lake, is a relatively easily observed phenomenon of turbulent diffusion (Langmuir, 1938; McLeish, 1968). Nevertheless, a very few experiments on the dispersion in the vicinity of singularities provided data that could be compared with the present theory. Among these few experiments, Csanady (1963) reported the time variations of group size of objects apparently in the presence of surface convergences. Reversal or suppression of turbulent diffusion was observed (see Figs. 4-6 of Csanady's paper). In one case, a group of objects, after fluctuating in their size, attained a limiting value. In another case of dispersion, the group size increased in the beginning and thereafter decreased to a small patch.

The theoretical results obtained also suggest that it may be possible from analyses of the dispersion of floatables to explore the characteristics of the velocity field in frontal zones.

Molinari and Kirwan (1975), Okubo and Ebbesmeyer (1976) and Okubo, Ebbesmeyer and Helseth (1976) have recently developed a method to determine the deformation rates, vorticity and divergence (or negative convergence) as well as eddy diffusivities from analyses of drogue observations. The method is based on expanding the velocity components of each drogue about the centroid of the drogues. The higher order terms of the expansion beyond the linear terms are regarded as "turbulence". The use of linear regression procedures enables one to obtain the mean velocity, deformation rates, vorticity, divergence, and eddy diffusivities. It would be interesting to apply this method to investigate precise flow patterns and turbulence in frontal zones.

Critical Size of a Growing Population, e.g. Phytoplankton Bloom Within a Convergence Zone.

Phytoplankton patches are often found in frontal zones (Bainbridge, 1957). Doubtless the high concentration of these organisms is partly attributed to the presence of convergence.

When physiologically unsuitable water surrounds a growing population, the community requires a minimum critical spatial extent to survive against dissipation by diffusion into the ambient water. Skellam (1951) and Kierstead and Slobodkin (1953) independently derived a theoretical critical size of a diffusing population with exponential growth.

They obtained a critical patch size $L = \pi(D/k)^{\frac{1}{2}}$ below which the patch could not be maintained and would be dissipated by diffusion (D is the [constant] diffusivity $\left[cm^2\ sec^{-1}\right]$ and k the phytoplankton growth constant $\left[sec^{-1}\right]$). For patch sizes greater than L, however, the patch would continue to grow indefinitely.

Okubo (1972) extended this theory by considering exponential cell growth in the presence of both surface diffusion and a steady convergence towards an attractive center. He found that the critical patch size L becomes $L = 2\lambda(D/k)^{\frac{1}{2}}$ where λ is the smallest positive root of the equation:

$$\tan\left[(1-m)^{\frac{1}{2}}\lambda\right] = \pm(m^{-1}-1)^{\frac{1}{2}} \qquad (6)$$

where $m = v^2/4kD$ and $v\left[cm\ sec^{-1}\right]$ is the convergence velocity. [The positive and negative signs in equation (6) refer $v > 0$ (attraction) and $v < 0$ (repulsion), respectively.]

The results of this theory are shown in Fig. 2. For convergence ($v > 0$), the

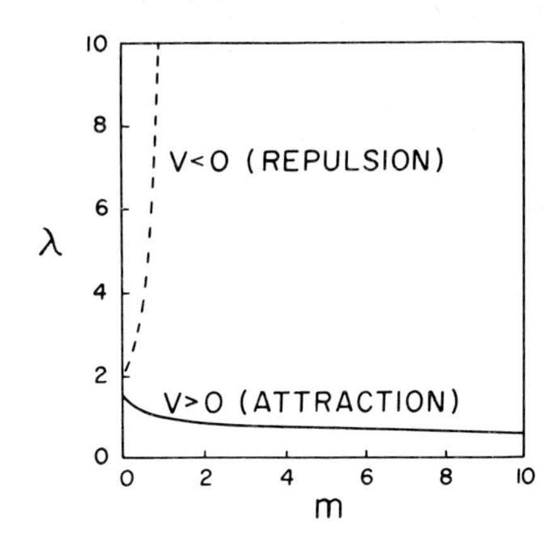

Fig. 2. Critical patch size as a function of velocity towards ($v > 0$) and away ($v < 0$) from a center of attraction (Okubo, 1972). λ is proportional to critical patch size and m is proportional to velocity^2.

critical patch size, as expected, becomes smaller as $m(\propto v^2)$ increases, and a narrow belt of phytoplankton results. As $m \to 0$, the solution collapses to $L = \pi(D/k)^{\frac{1}{2}}$, i.e., the Skellam-Kierstead-Slobodkin result.

However, in order for the patch to exist during active divergence ($v < 0$), the critical mass required increases rapidly, and for $m \geq 1$, i.e., $v^2 \geq 4kD$, becomes infinite, and a patch can never exist.

For example if $D = 10^5\ cm^2\ sec^{-1}$, $k = 1\ day^{-1}$, and $v = 0$, then $L \sim 1000m$, and if $v \leq -2cm\ sec^{-1}$, then $L \to \infty$. Hence, even a weak divergence of a few centimeters per second is capable of dispersing the strongest concentration of cells.

References

Bainbridge, R. 1957. The size, shape and density of marine phytoplankton concentrations. Cambridge Philosophical Soc., Biol. Rev., 32, 91-115.

Bowden, K. F. 1965. Horizontal mixing in the sea due to a shearing current. J. Fluid Mech., 21, 83-95.

Csanady, G. T. 1963. Turbulent diffusion in Lake Huron. J. Fluid Mech., 17, 360-384.

Kierstead, H. and L. B. Slobodkin. 1953. The size of water masses containing phytoplankton blooms. J. Mar. Res. 12, 141-147.

Klemas, V. and D. F. Polis. 1977. Remote sensing of estuarine fronts and their effects on pollutants. Phytogrammetric engineering and remote sensing, 43 (5), 599-612.

Langmuir, I. 1938. Surface motion of water induced by wind. Science, 87, 119-123.

McLeish, W. 1968. On the mechanisms of wind-slick generation. Deep-Sea Res., 15, 461-469.

Molinari, R. and A. D. Kirwan. 1975. Calculations of differential kinematic properties from Langrangian observations in the western Caribbean Sea. J. Phys. Oceanogr. 5, 361-368.

Okubo, A. 1970. Horizontal dispersion of floatable particles in the vicinity of velocity singularities such as convergences. Deep-Sea Res., 17, 445-454.

Okubo, A. 1972. Note on small organism diffusion around an attractive center: Soc. Japan, 28, 1-7.

Okubo, A., C. C. Ebbesmeyer, J. M. Helseth and A. S. Robbins. 1976. Reanalysis of the Great Lakes Drogue Study data. Special Report No. 2, Reference 76-2, Marine Sciences Research Center, State University of New York at Stony Brook, 84 pp.

Okubo, A., C. C. Ebbesmeyer and J. M. Helseth. 1976. Determination of Lagrangian deformations from analysis of current followers. J. Phys. Oceanogr. 6, 524-527.

Okubo, A. and C. C. Ebbesmeyer. 1976. Determination of vorticity, divergence and deformation rates from analysis of drogue observations. Deep-Sea Res., 23, 349-352.

Skellam, J. G. 1951. Random dispersal in theoretical populations. Biometrika, 38, 196-218.

5. SHALLOW SEA FRONTS PRODUCED BY TIDAL STIRRING

JOHN H. SIMPSON AND ROBIN D. PINGREE

1. Introduction

The European continental shelf is an area of generally large tide with associated currents frequently exceeding 1m sec^{-1}. These currents in the relatively shallow water of the shelf seas (Fig. 1) induce high levels of turbulent dissipation.

It has been estimated that approximately 10% of the total global dissipation of the M_2 tide in shallow seas occurs on the northwest European continental shelf. Vertical stirring associated with such high dissipation levels is sufficient in some areas to mix downward the seasonal buoyancy input at the surface and prevent the formation of stratification. The boundary between the vertically mixed and Stratified regimes frequently is delineated by a well defined front with a sharp change in sea surface temperature. We

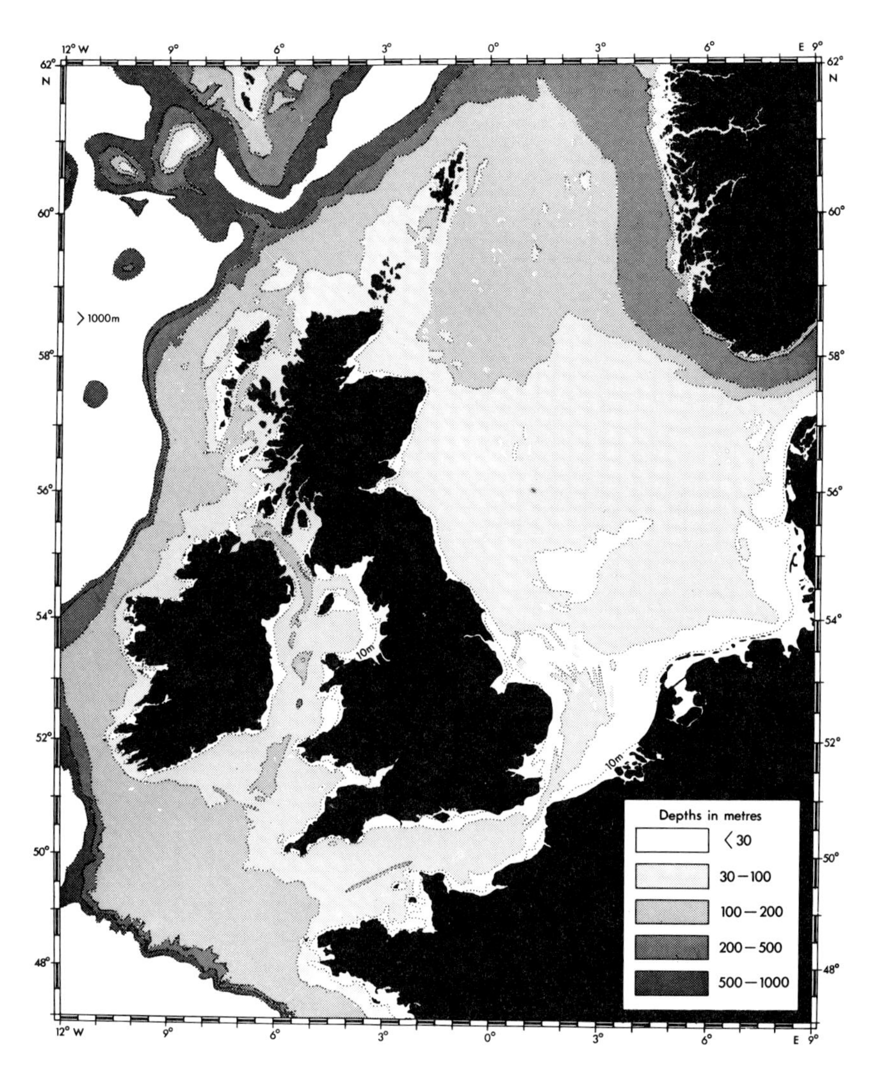

Fig. 1. Bottom contours of the continental seas around the British Isles and western Europe.

denote such zones as shallow sea fronts.

2. Conditions for Stratification

A convenient measure of the tendency for the water column to stabilize under the influence of a surface heat flux, Q, is the ratio R of the rate of production of potential energy necessary to maintain vertically well mixed conditions, to the rate of tidal energy dissipation; viz.,

$$R = \{g\alpha Qh/2C_p\rho\}/\{C_d|u|^3\} \propto \{h/|u|^3\},$$

if Q is constant, where α is the volume coefficient of expansion, C_p is the specific heat at constant pressure, ρ is the density, C_d is the drag coefficient, h is the water depth and $|u|$ is the amplitude of the tidal stream.

Contours of this Simpson-Hunter (1974) stratification parameter h/u^3 (Fig. 2) can be derived from the available tidal stream data or perhaps more satisfactorily, from a numerical model (Fig. 3). (Only the shape of the contour should be considered in comparing these figures, since the units of the plotted quantities differ.)

The high values of h/u^3 in the Celtic Sea, to the west of Scotland (Fig. 2) and in an isolated maximum in the western Irish Sea (Fig. 2) are associated with well stratified conditions in the summer months; low values in the English Channel, Bristol Channel and much of the Irish Sea coincide with regions that are well mixed vertically throughout the year. Infra-red images (Figs. 4 and 5) and analysis of ship observations (Fig. 6) indicates that frontal regions between well mixed and stratified conditions do indeed reflect the shape of these contours.

Figure 6 is a composite picture based on a static stability parameter, defined by

$$\overline{V} = 1/h \int_{-h}^{o} (\rho - \overline{\rho})\, gzdz$$

for August observations over the last 50 years. $\overline{V}$ is the amount of energy input required per unit depth to bring about complete vertical mixing of a formerly stratified water column with density distribution $\rho(z)$;

$$\int_{-h}^{o} \rho dz = h\overline{\rho} \quad .$$

3. Circulation Within Shelf Frontal Zones

At the present time we have only limited information on the longitudinal and transverse motion associated with shelf sea fronts. In view of their large alongfront scale (~100km) it is tempting to interpret the density field in terms of geostrophically balanced flow along the front. This simple approach predicts the existence of a jet within the strongly baroclinic zone with speeds of ~30 cm sec^{-1} parallel to the front. Direct observations of residual velocities in frontal regions, however, have not confirmed this. Rather, observations indicate currents in the range 2-16 cm sec^{-1} in directions which are often, but not always, parallel to the mean direction of the fronts. Remotely sensed infra-red images of the front show evidence of large scale instabilities and suggest that these fronts are not quasi two-dimensional structures with a well defined longitudinal flow.

Figure 4 is a sequence of three images spaced one day apart taken in August 1976. A marked meandering of the front with a wavelength of ~25 km is evident on August 18th. In the two days following, the meanders are seen to grow and pinch off, and two cold water eddies (lighter shade) become isolated in the stratified zone (darker shade). By a comparison of the successive images, the characteristic velocities of these instabilities are found to be ~10 cm sec^{-1}. Evidence for similar convolutions is found in aircraft IR and ship surveys of these fronts.

Our knowledge of the vertical circulation is even more limited and is based only on indirect approaches. A region of convergence at the surface may sometimes be inferred from the strong

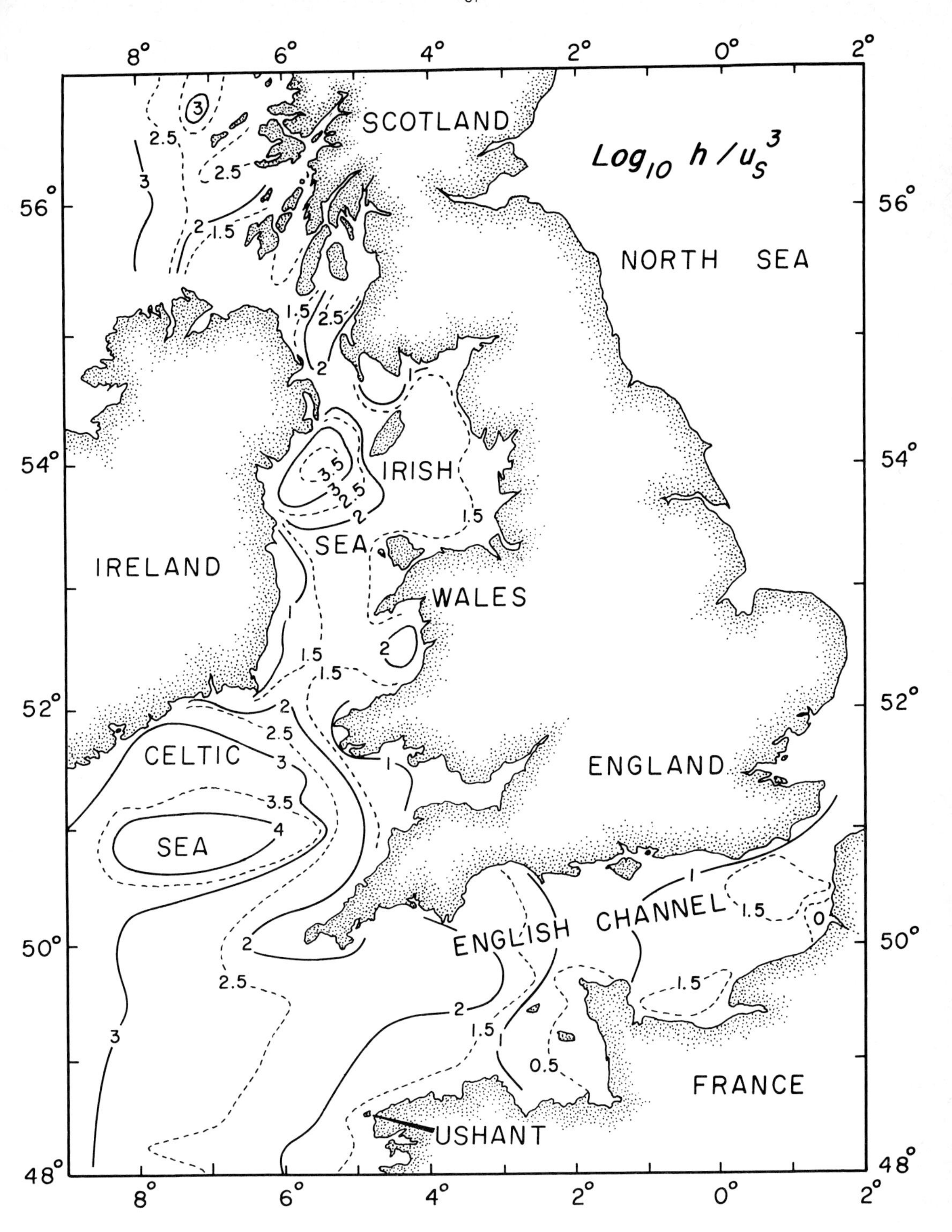

Fig. 2. The stratification parameter $\log_{10} h/u^3$ where h = water depth in meters and u = surface tidal stream amplitude in m $\sec^{-1}$ at mean spring tides.

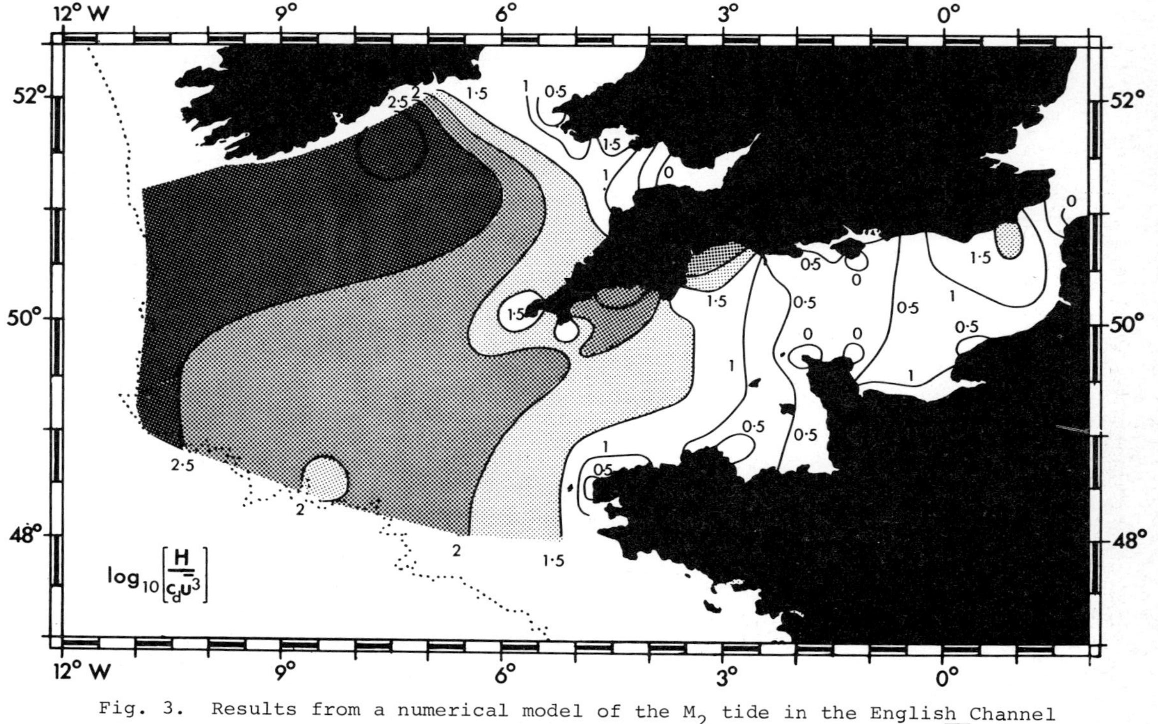

Fig. 3. Results from a numerical model of the M_2 tide in the English Channel and Celtic Sea of the stratification parameter $\log_{10}$ $(H/C_d\ \overline{u^3})$, where H is the water depth (cm) and $C_d\ \overline{u^3}$ represents the mean dissipation rate of M_2 tidal energy (cgs units).

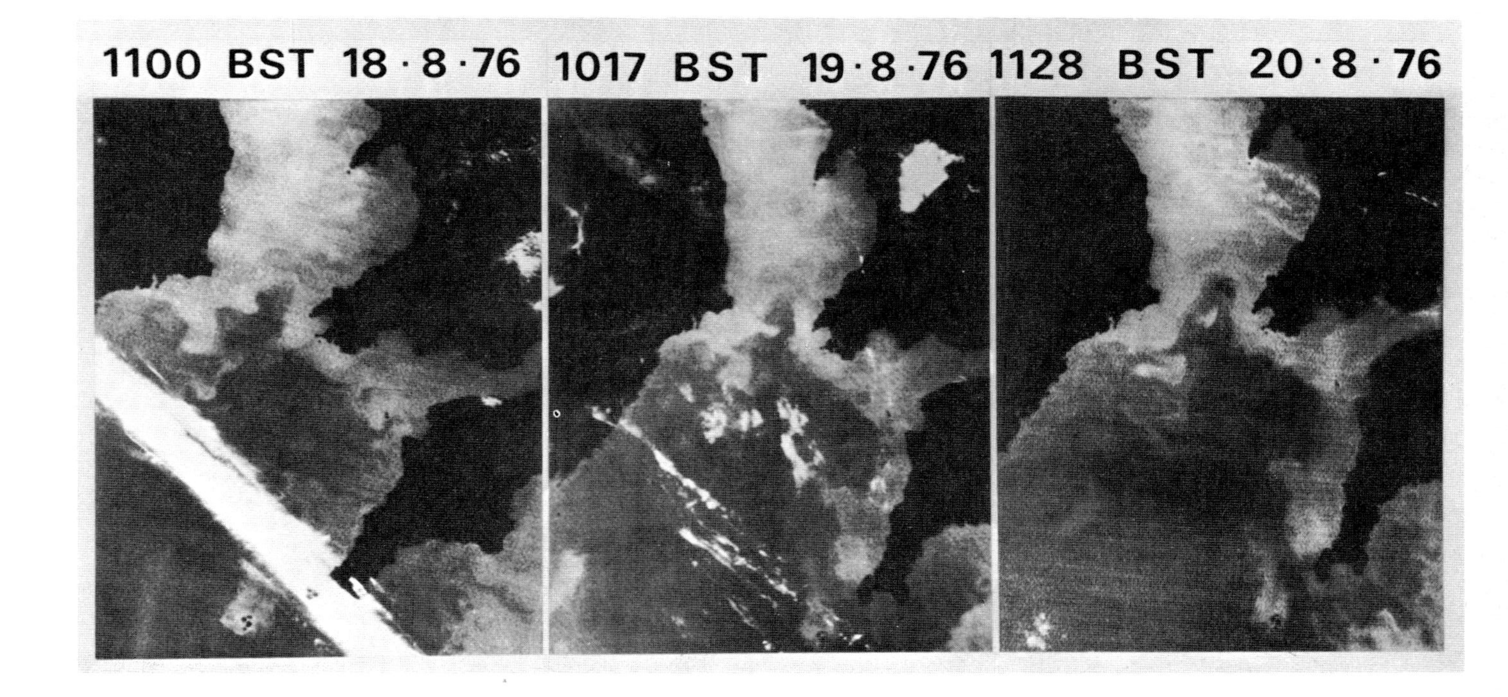

Fig. 4. Infra-red images from the NOAA 5 satellite. Sequence of pictures of the Celtic Sea front in the period 18-20 August 1976.

Fig. 5. Image of the Celtic Sea, Irish Sea and English Channel for August 26, 1976.

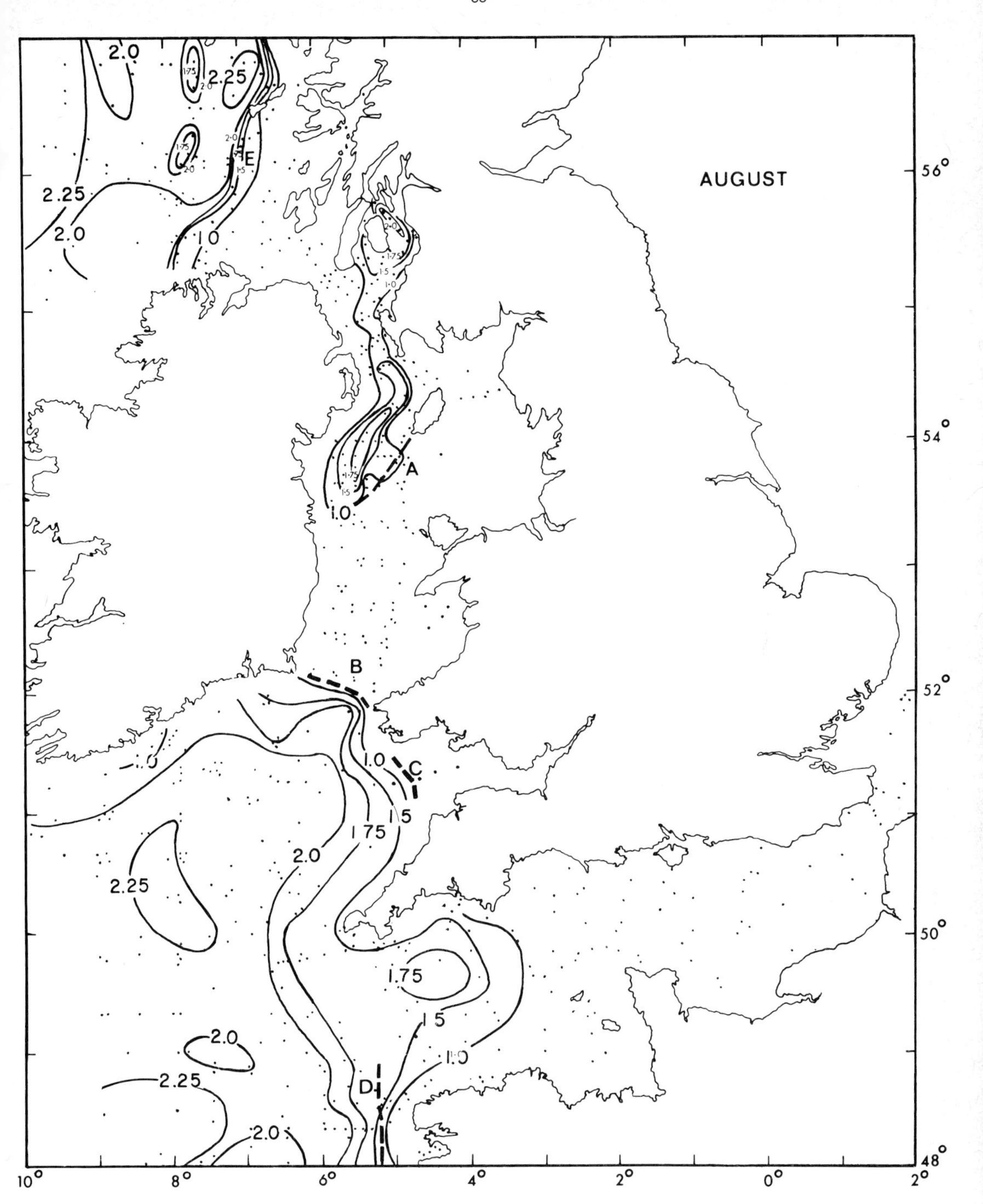

Fig. 6. Contours of the stability parameter $\log_{10} \overline{V}$ deduced from August ship of opportunity data. A, B, C, D, E, represent regions where fronts have been observed.

accumulation of material which occur in lines along the direction of the front and sometimes in the dense phytoplankton blooms that are displaced away from the frontal boundary toward more stratified regions (Fig. 7). At the same time there are frequently puzzling indications of areas of anomalously cold water in the vertically mixed water just before a sharp temperature rise across the front into the stratified side, and some upwelling may be taking place.

4. Chlorophyll *a*, Surface Temperature and Nutrient Kinetics.

Numerical models, infra-red satellite images and *in situ* measurements of temperature structure provide the necessary hydrographic background for interpreting the chlorophyll *a* distributions that are associated with frontal regions. The frontal characteristics in the Ushant (Ile d'Ouessant; off the northwest coast of France) frontal system (Fig. 2) are particularly well developed with sea surface temperature contrasts of 4-5°C (Fig. 8) found between the well mixed, strong tidal conditions near the French Coast and the more stable, weaker tidal regions of the Celtic Sea.

Repeated chlorophyll *a* traces through this frontal system have defined a region on the stratified side of the frontal boundary where dense phytoplankton blooms may persist throughout the summer months (Fig. 9). On the unstratified side of the frontal region strong tidal stirring ensures complete vertical mixing and the plant cells in the water experience a much lower mean light level than those held in the surface layers on the stratified side of the frontal boundary. Whereas in the well mixed region, levels of inorganic nutrients are relatively high, in the stable regions, surface nutrient values have remained low since being depleted in the spring bloom (Fig. 10). Thus the frontal zone is a region where the combination of high nutrients and a non-limiting light regime create conditions suitable for rapid or sustained phytoplankton growth.

The limits of frontal movement in response to the neap to spring cycle of tidal stirring define an area where phytoplankton populations tend to be dispersed at spring tides but experience conditions suitable for growth as the water column stabilizes at neap tides (Fig. 11). The eastern inshore boundary of a more persistent bloom is marked by the position of the front just after spring tides.

Its western outer limit is controlled by a deeper and more stable thermocline through which there is little or no vertical exchange between the surface and bottom mixed layers and consequently, in this region, production at the sea surface is nutrient limited. However, production may still occur in the thermocline just above the bottom mixed layer. During spring tides turbulent bursting phenomena will inject inorganic nitrate and phosphate into the base of the thermocline rather than to the surface along the frontal boundary. Subsequent utilization by dinoflagellates may result in chlorophyll rich thermocline layers.

In the spring, diatoms are abundant, whereas in the summer dinoflagellates predominate. Unlike diatoms, dinoflagellates do not have a requirement for silicate and this is reflected in differences in the concentration gradients for nitrate and silicate through the thermocline. The traces in Fig. 12 were obtained by sampling at a depth corresponding to the mean depth of the thermocline and the chlorophyll-rich layer (25 m). It was then possible to measure levels of nitrate and silicate at various positions across the temperature gradient as the passage of internal waves (periods ~5-10 min.) caused slow up and down movement of the thermocline. Figure 12 shows that while there is approximately a

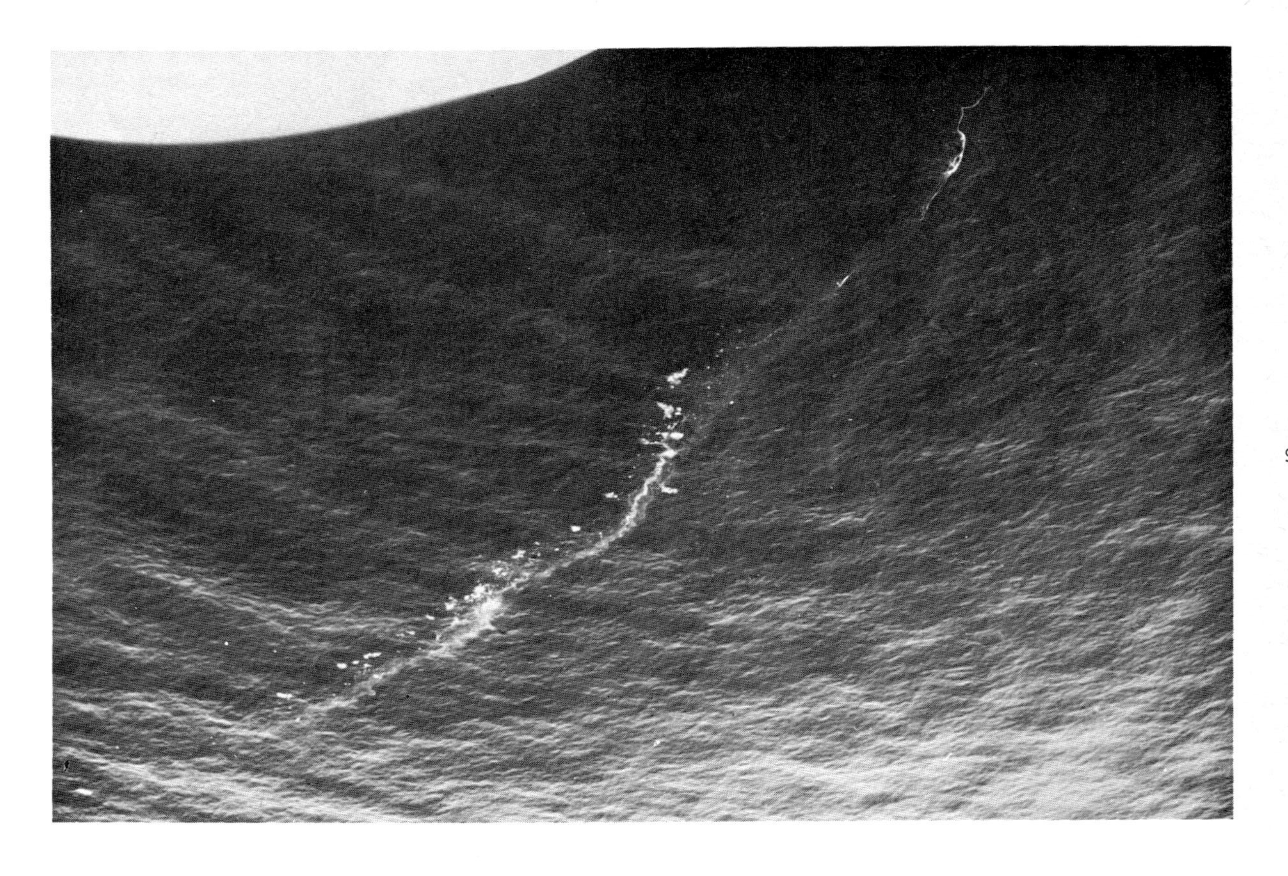

Fig. 7. Surface scum at 49°00' N, 5°30' W, August 9, 1976, from 100 m (see Fig. 11d).

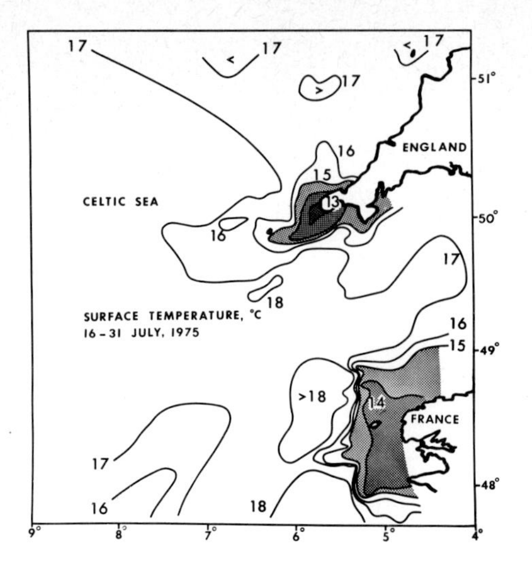

Fig. 8. Sea surface temperature (C), 16-31 July 1975, in the western approaches to the English Channel. Note the tendency for the frontal regions (as identified by 15 C temperature contour) to follow the stratification parameter contours (Figs. 2 and 3).

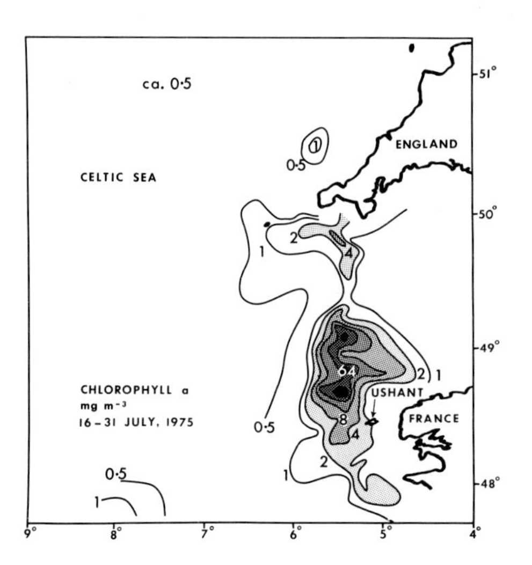

Fig. 9. Surface distribution of chlorophyll *a* (mg m^{-3}; 16-31 July 1975) corresponding to the frontal structure of Fig. 8.

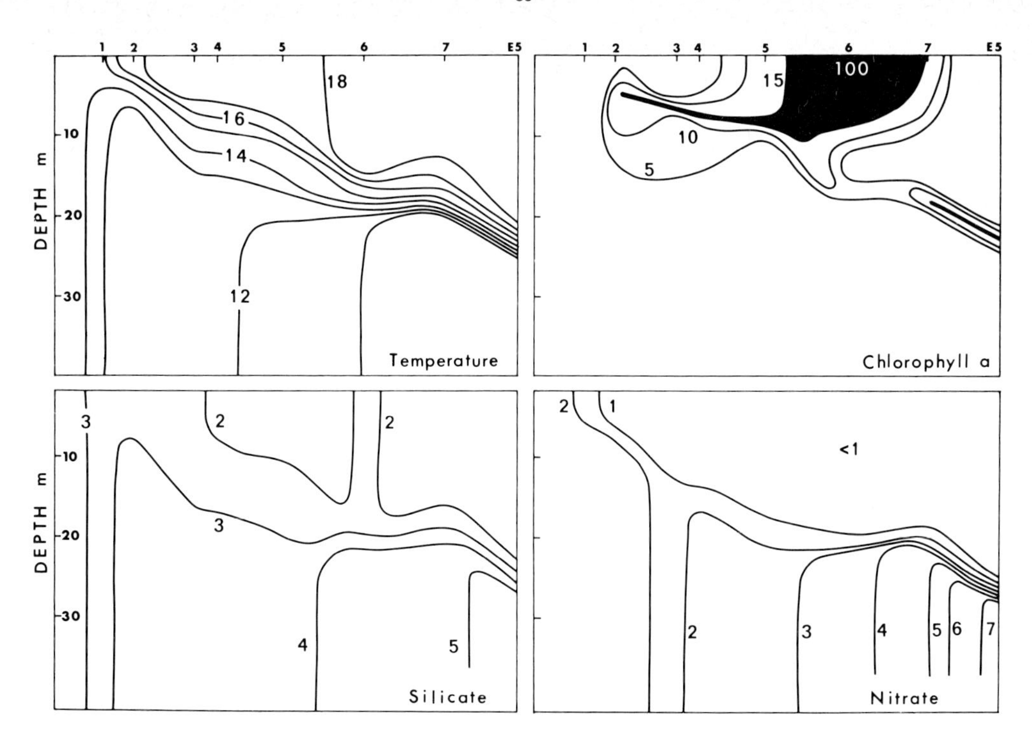

Fig. 10. Vertical sections for temperature (^{O}C), chlorophyll *a* (mg m^{-3}), silicate and nitrate (μg atom ℓ^{-1}) from mixed (station 1) to well stratified water (station E5) for August 10, 1976. The station positions are shown in Fig. 11a.

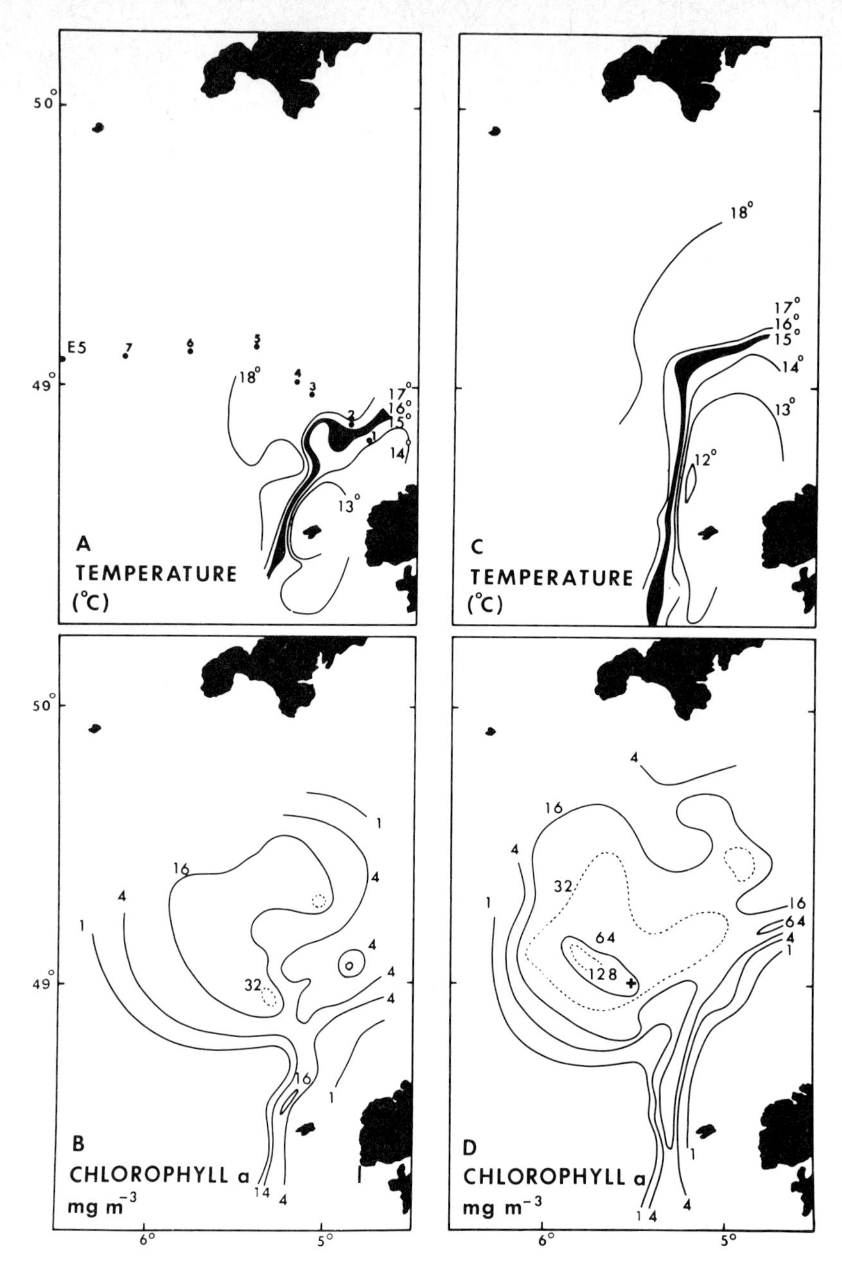

Fig. 11. Response of the Ushant frontal system to the spring-to-neap cycle of tidal stirring:

a) The frontal position under stabilizing conditions during neap tides relaxes back towards the French Coast (July 27-28, 1976).

b) Corresponding chlorophyll *a* distribution (mg m^{-3}).

c) The stratified region is eroded with the release of nutrient rich cold water at the surface under conditions of increasing turbulent energy production during spring tides (July 31 - August 2, 1976).

d) Corresponding chlorophyll *a* distribution (mg m^{-3}).

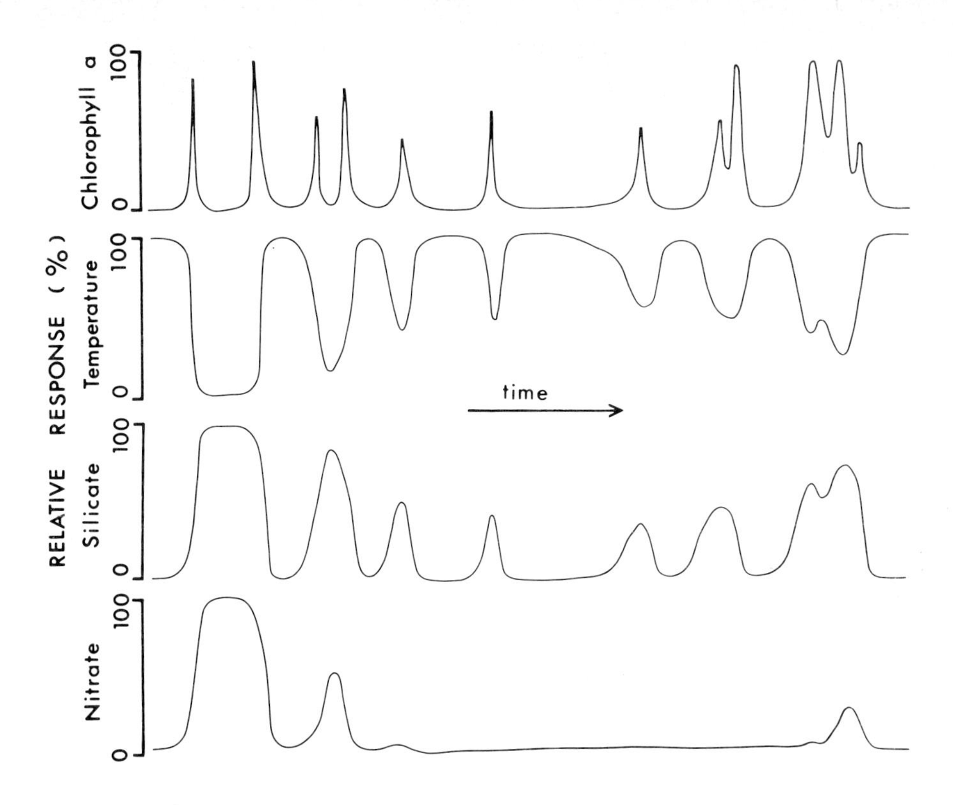

Fig. 12. Records from station E5 (Fig. 10) (1450-1610 GMT August 10, 1976) for nitrate, silicate, temperature and chlorophyll *a* at the mean depth of the thermocline (25 m), showing the effects of internal waves propagating past the pump intake. Each is presented on a scale of percentage change between values for the surface and bottom mixed layers.

one to one correspondence between the temperature and silicate traces, the nitrate values increase significantly only below the chlorophyll peaks.

Silicate, therefore, mixes as a conservative property. On the other hand, nitrate is utilized quickly after its release by tidal stirring and subsequent stabilization. Continual dinoflagellate growth in the thermocline and at the surface in the frontal region is, therefore, dependent upon nutrient release through tidal stirring and its availability through stabilization of the water column by heat penetration downwards.

References

Fearnhead, P. G. 1975. On the formation of fronts by tidal mixing around the British Isles. Deep-Sea Res., 22, 311-321.

Pingree, R. D. 1975. The advance and retreat of the thermocline on the continental shelf. J. Mar. Bio. Assn. U.K., 55, 965-974.

Pingree, R. D. 1977. The bottom mixed layer on the continental shelf. Estuarine and Coastal Marine Science 5, 399-413.

Pingree, R. D., P. M. Holligan and R. N. Head. 1977. Survival of dinoflagellate blooms in the western English Channel. Nature, 265, 266-269.

Pingree, R. D. 1976. The influence of physical stability on spring, summer and autumn phytoplankton blooms in the Celtic Sea. J. Mar. Biol. Assn., U.K., 56, 845-873.

Pingree, R. D., P. R. Pugh, P. M. Holligan, and G. R. Forster. 1975. Summer phytoplankton blooms and red tides along tidal fronts in the approaches to the English Channel. Nature, 258 (5537) 672-677.

Pingree, R. D., G. R. Forster, and G. K. Morrison. 1974. Turbulent convergent tidal fronts. Journal Marine Biological Assoc. of the U.K., 54, 469-479.

Savidge, G. 1976. A preliminary study of the distribution of chlorophyll in the vicinity of fronts in the Celtic and western Irish Seas. Estuarine and Coastal Marine Science, 4, 617-625.

Simpson, J. H. 1971. Density stratification and microstructure in the western Irish Sea. Deep-Sea Res., 18, 309-319.

Simpson, J. H. 1976. A boundary front in the summer regime of the Celtic Sea. Estuarine and Coastal Marine Science, 4, 71-82.

Simpson, J. H., and J. R. Hunter. 1974. Fronts in the Irish Sea. Nature, 250, 404-406.

Simpson, J. H., D. G. Hughes and N. C. G. Morris. 1977. The relation of seasonal stratification to tidal mixing on continental shelf. D. Sea Research, Deacon Birthday Issue.

6. PROGRADE AND RETROGRADE FRONTS

CHRISTOPHER N. K. MOOERS, CHARLES N. FLAGG
AND WILLIAM C. BOICOURT

Introduction

This section outlines some of the prominent features of the class of fronts that are located over the outer portions of the continental shelves and the upper portions of the continental slopes. These fronts are typically referred to as *upwelling* and *shelfbreak* fronts. The terms are somewhat misleading as they imply mutually exclusive dynamics, in fact, these fronts can co-exist. Their greatest distinctions may rest in the relative strengths of the large scale forcing, e.g. offshore surface Ekman transport versus alongshore transport of freshwater runoff, and wind stirring of the surface layer versus tidal stirring of the bottom layer. We suggest a more general terminology to describe these fronts might be *prograde* and *retrograde* fronts implying that the frontal isopleth slopes are the same or opposite to the cross-shelf topography, respectively. In Fig. 1 are sketched some simplified forms of frontal isopleths relative to the topography that are commonly encountered. Some combination of these fronts probably occur on all the continental shelves of the world.

These fronts are distinguished from other types of fronts such as river plume or estuarine fronts, shallow sea fronts, and mid-oceanic fronts by the hydrographic and biological space and time scales of the gross features and the strong bathymetric influences. The gross features of the fronts are synoptic in scale, that is, they have a cross-front scale of the order of the internal radius of deformation, R_{bc} ~30 km, and along-front scale of many radii of deformation and a persistence scale greater than several inertial periods. Thus, the alongshore flow is essentially in geostrophic balance. However, there are important ageostrophic (frictional and nonliner) flows on a scale smaller than the radius of deformation, especially in the cross-frontal flows.

Because of their location, and widespread occurrence, these fronts have appreciable significance for at least the general circulation of the coastal ocean and the air-sea interaction of the coastal zone. They represent a boundary across which exchange of momentum, heat, salt, organic and inorganic materials between the coastal and open ocean must occur. They may, thus, be a candidate zone for the seaward boundary of coastal circulation models. In any event, they will have to be monitored eventually; a subsidiary goal of the corresponding research will be to design monitoring strategies.

Fronts are typically influenced by large scale ($\gg R_{bc}$) geostrophic and Ekman flows which produce frontogenesis. On the cross-front scale of the R_{bc}, the alongfront flow is in geostrophic balance, typically in the form of relatively intense baroclinic surface and subsurface jets, and the cross-front flow is dominated by surface and bottom Ekman layer dynamics. At scales less than the R_{bc}, there are substantial ageostrophic flows, particularly in the cross-front flow, which may be driven by internal turbulent boundary layers.

These simple considerations provide some priorities in sampling schemes. For example, on the large scale or span, only geostrophic (density profiles and a few interior direct current and bottom pressure measurements for calibrating the absolute geostrophic velocity field) and Ekman (wind stress field and a few surface and bottom layer direct current measurements) information are needed, but over a season. With a geostrophic-Ekman diagnostic (hindcasting) model, the (2-day) low passed data can be interpolated to produce, say, weekly, 3-D velocity and density fields.

For the ageostrophic (frictional and

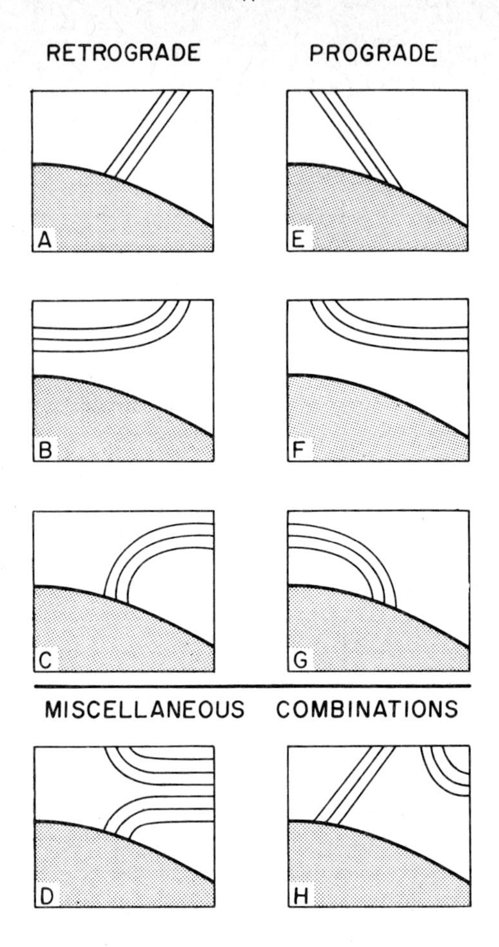

Fig. 1. Topology of prograde and retrograde fronts, with miscellaneous combinations.

A. Surface and bottom influenced retrograde front.
B. Surface layer influenced retrograde front.
C. Bottom layer influenced retrograde front.
D. Combination of types C and F.
E. Surface and bottom influenced prograde front.
F. Surface layer influenced prograde front.
G. Bottom layer influenced prograde front.
H. Combination of types A and F.

nonlinear) regime, sampling at cross-front intervals less than the R_{bc} is required. The vertical and temporal resolution requirements are correspondingly greater. The alongfront resolution requirements are more subtle because it is not yet clear whether along-front variability is predominantly phase-locked to bathymetry (submarine canyons and capes) or propagating, coastally-trapped, eddy-waves. In any such anisotropic situation, it is essential to have some 3-D information in order to know the orientation of the gradients.

Retrograde Fronts

A familiar example of a retrograde front is found along the outer continental shelf of eastern North America during the winter. The front in this area is the transition between the cold and fresher shelf water mass and the warmer and more saline slope waters further offshore. This shelf/slope water front, dominated by the same basic dynamics, extends from Cape Hatteras to the south, throughout the Middle Atlantic Bight, extending in some form northeastward to the transition region between the shallow waters of the Grand Banks and the Atlantic Ocean. A similar retrograde front also exists at times in the nearshore regions of the South Atlantic Bight, shoreward of the prograde front associated with the northern (nearshore) wall of the Gulf Stream.

In the following paragraphs is a discussion of the hydrography and some kinematic properties of the shelf/slope front in the Middle Atlantic Bight where this type of front is perhaps best documented. The early studies of Bigelow and Sears (Bigelow, 1933; Bigelow and Sears, 1935), the later work of Cresswell (1958), and the more recent surveys of Boicourt (1973, 1975), Beardsley and Flagg (1975), and Wright (1976) have defined most of the major features of the shelf/slope frontal hydrography in the Bight. Some of the drifter studies discussed by Bumpus (1973) and those of Voorhis, Webb, and Millard (1976), together with the direct current measurements of Beardsley and Flagg (1975) and Flagg (1977) have outlined mean current characteristics of the front and have started to define the variability of the current field.

Seasonal characteristics. The front has a simple structure during the winter months of November through March (Fig. 2). The water on the shelf is fairly homogeneous in the vertical, with temperatures and salinities increasing from 4°C and 32.5°/oo near shore to about 8°C and 33.5°/oo just shoreward of the front. The density structure is dominated by the salinity field so that sigma-t increases offshore from 25.5 near shore to 26.1 just inshore of the front. Slope water offshore of the frontal zone has surface temperatures of 14 to 16°C and surface salinities of 35.5 to 36.0°/oo.

Both temperature and salinity generally decrease with depth over the continental slope with water at 1000 m typical of North Atlantic Central Water at 4.0°C and 34.94°/oo. Thus across the front, temperature, salinity, and sigma-t change by approximately 5°C, 2°/oo, and 0.5 σ_T units, respectively. Hence, a characteristic property of the shelf/slope front is the strong tendency for the temperature and salinity gradients to have a compensating effect on density, which tends to minimize the density contrast. These changes occur in a horizontal distance ranging from 7 to 40 km and a vertical depth ranging from 15 to 60 m. The center of the front south of New England generally coincides with the 10°C isotherm, the 34.5°/oo isohaline, and the 26.5 sigma-t isopycnal .

A primary characteristic of the shelf/slope front in the Middle Atlantic Bight is that the front always seems to straddle the shelf break, the zone of abrupt increase in bottom slope marking the transition between the continental

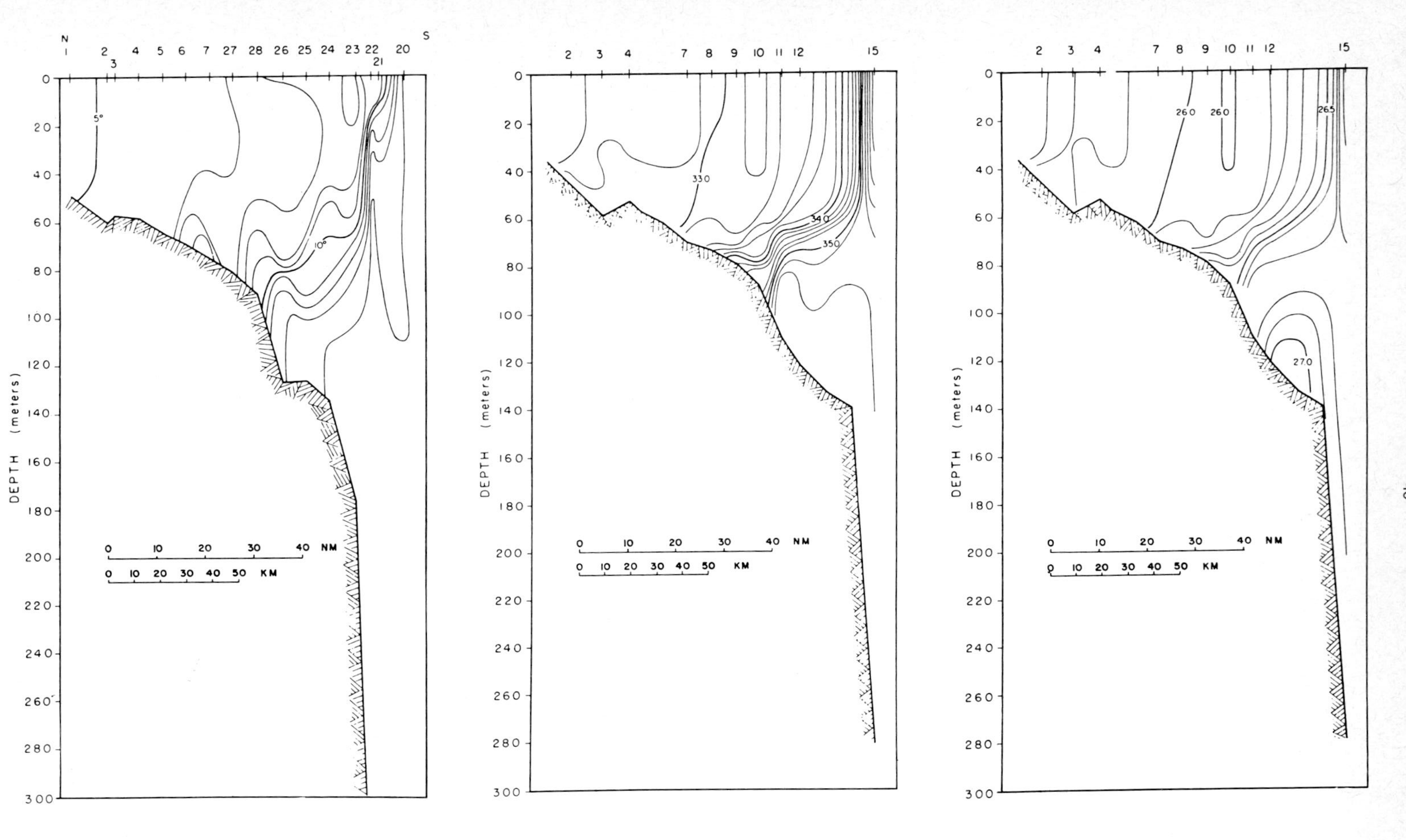

Fig. 2. Cross-shelf hydrographic sections, south of Rhode Island, April 2-4, 1974 (from Flagg and Beardsley, 1975).

shelf and upper continental slope. South of New England, the shelf break occurs at about 120 m, is somewhat deeper toward the east and somewhat shallower toward the west and southwest. The center of the front intersects the bottom generally near the 100 m isobath (Wright, 1976) but can range from the 60 m isobath to deeper than the 120 m isobath depending upon the individual section. The front slopes up from the bottom and offshore for a distance that ranges from 0 to 50 and perhaps as much as 100 km. A typical slope for the interface is approximately 2×10^{-3}, a value which is about twice the average continental shelf bottom slope and less than one tenth the maximum bottom slope of the underlying continental slope. South of New England the surface expression of the front seems to be generally confined within the 500 m isobath. The winter front is thus conceived simply as a narrow inclined region over moderately steep topography separating a somewhat vertically homogeneous shelf water mass from a denser and mildly stratified oceanic water mass.

The vernal progression drastically alters the simple winter frontal structure just described. Increasing solar radiation and diminishing storm frequency and intensity permit the formation of the seasonal thermocline over both the shelf and slope regions (Fig. 3). In the Middle Atlantic Bight during the late spring and summer a cold band of shelf water, generally delineated by the $10^{\circ}C$ isotherm, occupies the mid-shelf region below the seasonal thermocline. This band is surrounded (inshore, above, and offshore) by warmer waters. Thus, a well defined sub-surface temperature front remains below the thermocline as the offshore boundary of the cold band.

Increased fresh water runoff in the spring decreases the shelf water salinity slightly to its annual minimum in early summer, thus enhancing the salinity front even in the near surface waters. However, the seasonal thermocline dominates the density variations of the near surface water so that the seasonal pycnocline extends from the shelf well out into the slope waters and the sharp winter density front essentially disappears. Nevertheless, a surface temperature front persists through summer which, although it is usually not detected during hydrographic surveys, is sufficiently intense to be detected by satellite and aircraft IR imagery.

A weak density front is maintained at the offshore edge of the cold band below the seasonal pycnocline. The important point here is that although the front is altered, weakened, and sometimes disrupted by intrusions, the formation of detached parcels, and other advective processes which will be discussed later, a frontal zone continues to exist throughout the year and occupies essentially the same position relative to the shelf break.

Frontal Variability. The following discussion of the variability of the frontal zone is primarily based upon the wintertime measurements of Beardsley and Flagg (1975) and Flagg (1977). The frontal zone is very active in the internal wave band with the greatest concentration of energy in the zone of large static stability of the front itself. It appears that the presence of the sharp density gradient and the proximity of the surface and bottom invalidates the assumptions of linear internal wave theory in the frontal zone. The semidiurnal and and diurnal tides account for the largest percentage of horizontal current variance in the front. The semidiurnal tide appears to be barotropic in the front as well as on the shelf while the diurnal tide increases toward the bottom.

The semidiurnal tide is less clockwise polarized than the diurnal, and has tidal ellipses aligned approximately with the isobaths. Inertial energy is important on the shelf and relatively more so in the frontal region where it has the

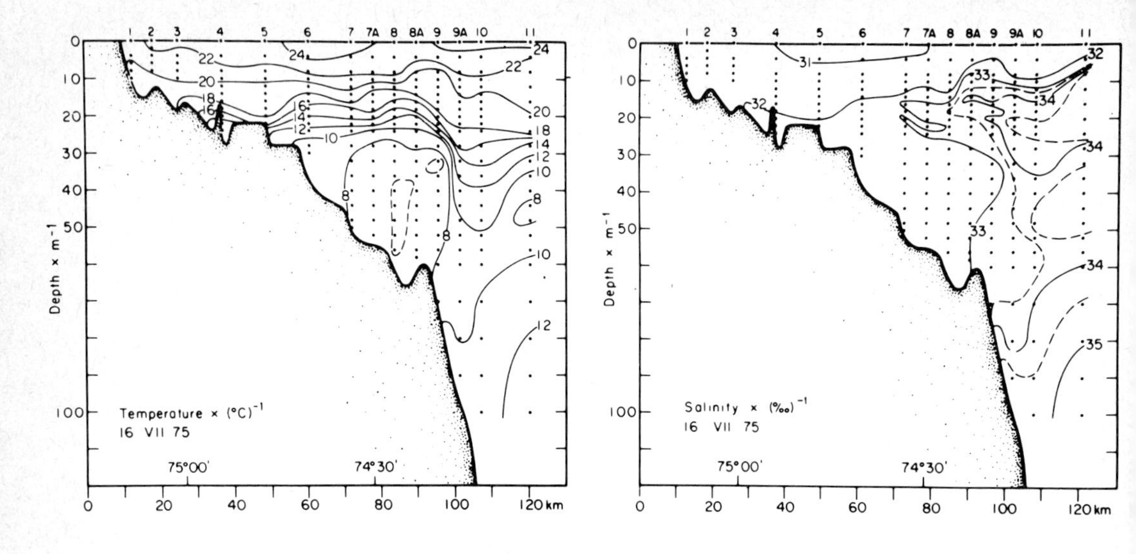

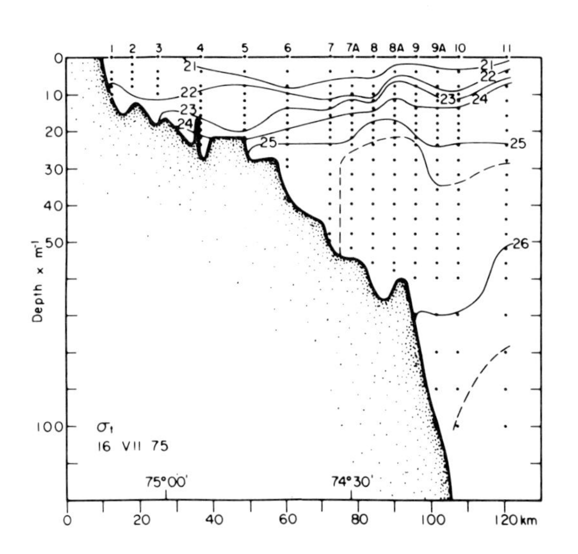

Fig. 3. Distributions of temperature, salinity, and σ_T in a cross-shelf vertical section off Ocean City, Maryland, July 1975. (from Beardsley et al, 1976)

same mean amplitude as the semidiurnal tide in the water above the front. The inertial energy decreases with depth as expected while the amplitude fluctuations are reasonably correlated with wind stress events. Maximum inertial current amplitudes can be three to four times the average.

At periods longer than a day, several frontal phenomena have been identified although details of their kinematics and dynamics are poorly understood in most cases. Satellite photos and some hydrographic surveys have shown that the front meanders about its mean alongshore position (Ingham, 1976; Beardsley and Flagg, 1975). One type of meander is typified by alongshore scales of 50 to 150 km, amplitudes of 10 to 50 km, and periods which are thought to range from approximately two days to two weeks. These meanders are confined to a region whose width increases approximately linearly from Cape Hatteras progressing north and eastward along the shelf edge (Fig. 4) being about 200 km wide south of New England. The "mean" position of the front exhibits an annual on and off shore variation within this region (Ingham, 1976).

The detailed vertical and horizontal structure of these meanders has not been well documented especially in the alongshore direction and their causes for the most part are not understood. A stability analysis of a front with density and current fields similar to those south of New England over a steeply sloping topography like that of the continental slope indicates that the front in the Middle Atlantic Bight, for practical purposes, is baroclinically stable (Flagg, 1977). Thus the meanders are not self induced by the frontal structure but must be forced by some outside influences.

Wind forcing. Wind stress has been shown to be the dominant forcing term for continental shelf currents at periods of two to ten days (Beardsley and Butman, 1974; Beardsley, et al., 1977; and Flagg, 1977). As such it is natural to expect that wind stress might also be a cause for frontal motions. Flagg (1977) showed that the currents in the layer above the front responded primarily to the alongshore component of the wind stress. The current generally responded with a large alongshelf flow, although not as large as at mid-shelf, and with a smaller cross-shelf circulation pattern consistent with an Ekman layer flow to the right of the wind with a return flow below. Evidence showed, however, that the responses varied in detail and seemed to be dependent upon the time history of the wind stress vector and the state of the frontal currents at the onset of high wind.

Flagg (1977) documented an interesting example of the response of the frontal region to wind stress forcing and clearly showed that consideration of frontal motion will be required for a complete understanding of the current response shoreward of the shelfbreak. In this instance, a strong northeastward wind produced a large Ekman offshore flow which forced the upper portion of the front offshore while a return flow along the bottom brought the lower portion of the front shoreward. The mid-depth part of the front seemed to stay fixed. After the passage of the storm, the front oscillated about its original position for ten or eleven days with a period of three to four days. The frequency of occurrence of this motion and its alongshore structure are unknown. Another type of response was observed by Boicourt and Hacker (1976) off the coast of Delaware during the summer of 1974 where an offshore Ekman flow forced a return flow in the form of a strong mid-depth intrusion of saline water 20 m thick pushing shoreward for approximately 60 km.

Advective influences. The shelf/

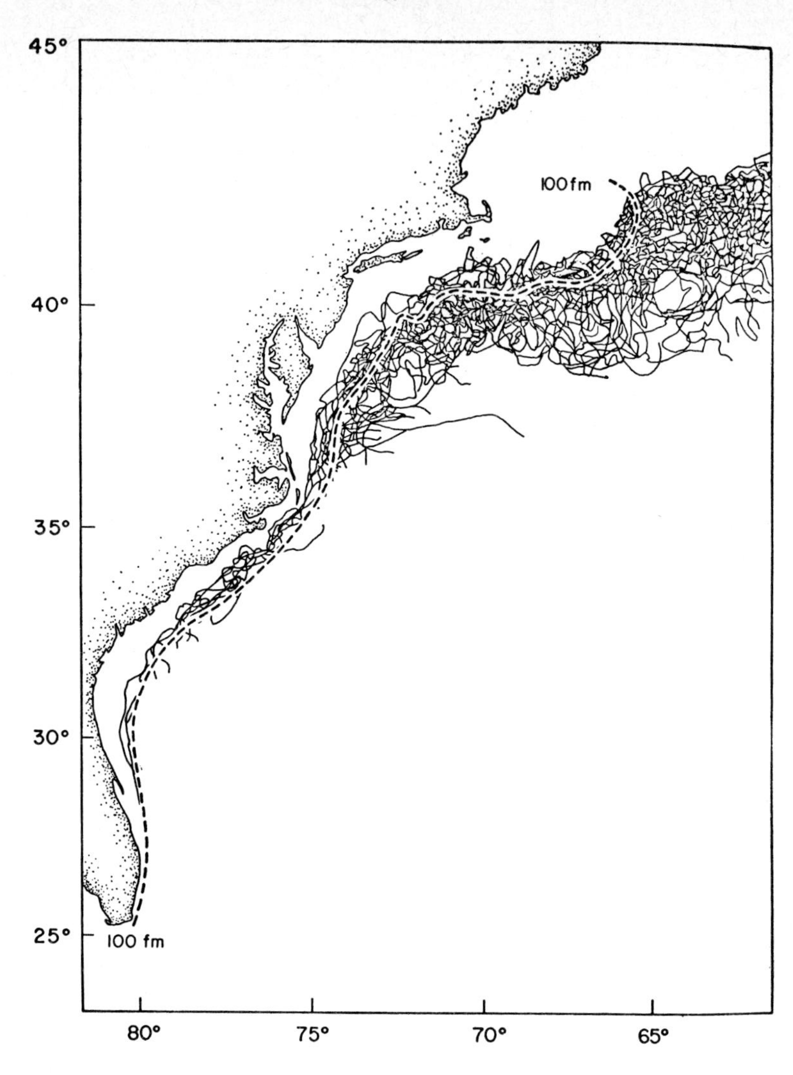

Fig. 4. Composite plot of the positions of the Middle Atlantic Bight shelf/slope front as observed by satellite from June 1973 to December 1974. Heavy dashed line indicates the position of the shelf break. (after Ingham, 1976)

slope front of the Middle Atlantic Bight is subject to two types of advective motions, both of which have been well documented as to their existence but whose detailed structures remain unresolved. The first of these is the calving process first described by Cresswell (1958) and well summarized as to its frequency of occurrence and cross-shelf character by Wright (1976). Calving refers to the process whereby lens-shaped parcels or "bubbles" of shelf water are transferred through the shelf/slope front into the slope water. These parcels show up in the cross-shelf sections of temperature and salinity as closed contours with cross-shelf widths on the scale of 10 to 20 km and thickness of 20 to 50 m offshore the front (Fig. 5). Often the at all depths down to the depth of the intersection of the front with the bottom.

Another characteristic of the calving is that the "bubbles" are not visible in the density contours although the paths of seaward migration seem to be along isopycnal surfaces. While the occurrence of the "bubbles" is apparently possible at all times of the year, there appears to be an annual cycle with the maximum frequency of occurrence generally in August and September.

Three primary questions remain about the calving process. The first is that the alongshore structure of the "bubbles" remains little more than speculation. Present concepts indicate an alongshore length at least comparable to the cross-

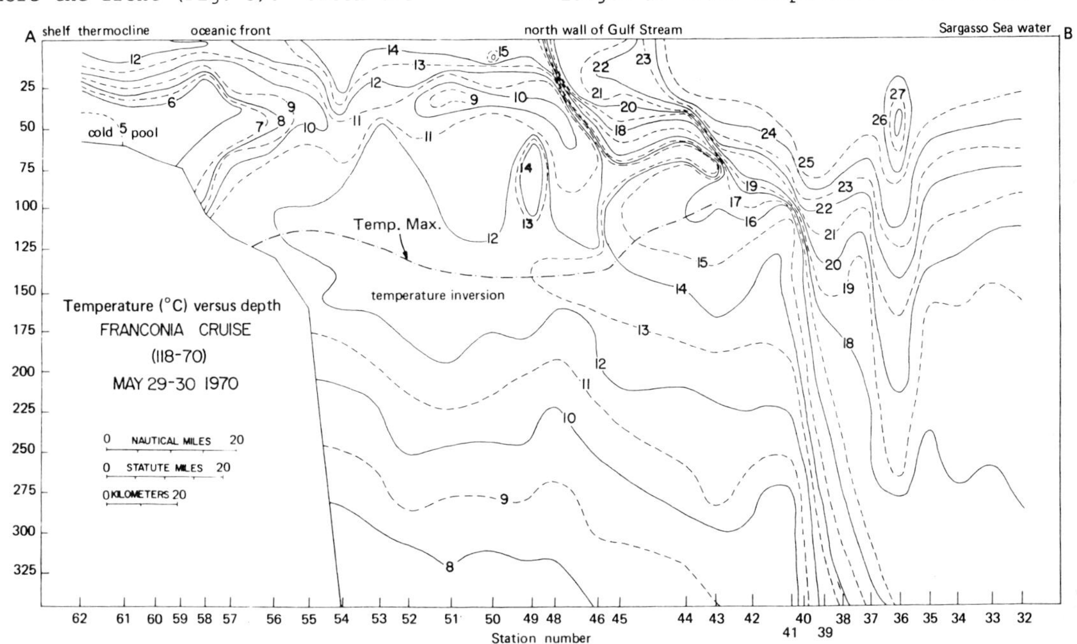

Fig. 5. New York Bight cross-shelf temperature section. A bubble of shelf water (T < 10C) can be seen between stations 46 and 52, and centered at ~40 m depth (from Bowman and Wunderlich, 1977).

cross-frontal section exhibits offshore distortions of the isopleths of temperature and/or salinity which were clearly at one time connected to the "bubbles" of shelf water. The "bubbles" have been seen shelf width, but there has been no observational confirmation of this. Secondly, the speeds of the seaward migration of the "bubbles" are unknown although attempts have been made with various kinds of drogues to measure them. The speeds of migration together with the physical dimensions of the "bubbles" are important for calculating the relative effectiveness that the calving process has on the total cross-frontal exchange.

The last question is perhaps fundamental in resolving the previous two. That is, what is the process that instigates the formation of the "bubbles" and their offshore migration? It will probably take a fairly large scale and operationally flexible experiment to resolve these questions.

The second type of advective motion affecting the front is that associated with the impingement of warm core eddies from the Gulf Stream onto the continental slope. Aside from the Gulf Stream and shelf/slope fronts, these eddies are the major features of infrared satellite images from the Middle Atlantic Bight area. Hydrographic sections through eddies indicate that they extend for several hundred meters below the surface, that there are large ($\sim$100 cm sec^{-1}) anti-cyclonic (clockwise) velocities, and that, as indicated by satellite photo information, the horizontal scales are of the order of 100 km.

The effects of the eddies on the shelf and front have been addressed by Saunders (1971), Morgan and Bishop (1976), and Chamberlin (1976). The eddies move along the shelf edge west and south at 0 to 10 cm sec^{-1} south of New England and they entrain shelf water into their northeast quadrant. One of the major areas of contention about the eddies is the amount of water involved in the entrainment. Once a clearer idea of the volume of the entrained tongues is obtained, the ease with which information can be obtained on the distribution and size of the eddies from satellite photos will yield much information on the importance of the eddies for the cross-frontal exchange budget.

There is a complementary question: while anticyclonic eddies entrain surface shelf waters, can the bottom shelf waters simultaneously entrain slope or Gulf Stream water through intrusion, etc. In addition to the entrainment there undoubtedly is some momentum, vorticity and energy transferred to the shelf through the front by the eddies. However, no studies have been addressed to this problem.

Implications for fisheries. The envelope of the Middle Atlantic Bight shelf/slope water front corresponds to the position of the great fishing grounds (for benthic and pelagic fish) running along the shelfbreak from the Grand Banks to Cape Hatteras (see Figs. 1 and 2, section 8). Considering the role that frontal circulation can play in the distribution of nutrients, phytoplankton, zooplankton, etc., this correspondence seems likely to be more than mere coincidence. C. Mooers has learned from Jacob J. Dykstra, President, Point Judith Fishermen's Cooperative Ass'n, Rhode Island, that fishing is best at the shelfbreak when the bottom turbidity is decaying after a great increase. This is probably due to organic particulate matter being put into suspension and providing food for zooplankton who are, in turn, fodder for fish. Situations of storm-induced, nearbottom feeding frenzies are valued by fishermen. They are so important that fishermen spend most of their time waiting for them to happen. Hence, a predictive capability for such situations would have significant economic value for fisheries.

Prograde Fronts

Coastal upwelling fronts are a familiar example of prograde fronts. The best known coastal upwelling fronts are those generally found in the subtropical latitudes along the west coasts of continents. These fronts owe their existence to the wind-driven coastal Ekman divergence, or coastal upwelling. Other examples of prograde fronts are those associated with western boundary currents where those currents are topographically trapped, again in subtropical latitudes, along the east coast of continents; e.g., the Gulf Stream in the Florida Straits and South Atlantic Bight. Wind-driven coastal upwelling occurs along these fronts, too, on a seasonal and a storm transient basis. These fronts do not owe their fundamental existence to local wind-driven upwelling but, rather, they are baroclinic features of the large scale, general oceanic circulation established by the invariance of potential vorticity and geostrophy. As such, their ultimate source may well be large scale open ocean, wind-driven upwelling. The present discussion will focus on coastal upwelling fronts *per se*, though many of the points mentioned carry over to western boundary current fronts.

It has long been known that sustained, wind-driven coastal upwelling occurs along the west coasts when the meridional winds are persistently equatorward, which induces an offshore Ekman transport. The offshore Ekman transport produces a mass deficit at the coastline, the so-called coastal Ekman divergence. Thus an upwelling of subsurface, offshore water into the surface layers nearshore is required for mass compensation, unless an alongshore convergence provides the mass compensation.

A wind pattern favoring upwelling occurs generally on a seasonal basis as the atmospheric subtropical high pressure zone intensifies over the ocean and moves poleward in summer. As a consequence, there may be at least weak upwelling year-round in low latitudes, but only seasonal upwelling in mid-latitudes, which spreads poleward as the summer season advances. The poleward edge of the subtropical high constitutes a boundary between the so-called Westerlies and the Easterlies, and is consequently a locus for storm tracks; i.e., atmospheric cyclones and anti-cyclones. Correspondingly, the poleward extremity of the seasonal coastal upwelling zone is marked by low persistence or high intermittency. This intermittency results in transient coastal upwelling or so-called coastal upwelling wind event cycles.

General characteristics. As noted above, the coastal upwelling process introduces subsurface (ca. 100 to 200 m or more), offshore (beyond the shelfbreak) water into the surface layer (upper 10 to 50 m) nearshore (basically within a baroclinic radius of deformation from the coastline; however, this distance is affected by buoyancy and momentum mixing). Thus, the upwelling process results in an upward tilt of density surfaces in moving from offshore to over the continental shelf. These tilted isopycnic surfaces result in strong baroclinicity (as great as that of the Gulf Stream). As the elevated isopycnals approach the surface nearshore, a hyperbaroclinic zone is frequently encountered; this is termed the frontal zone.

Under conditions of vigorous, sustained upwelling-favorable winds, the upwelling frontal zone outcrops at the sea surface to form a strong surface front (there are more subtleties to this process which will be addressed below). Due to the extreme baroclinicity of the frontal zone, it is, on average, not far from dynamic instability. Superposition of vertical shears due to internal tides or near-inertial motions can drive it dynamically unstable. Similarly, enhanced upwelling; i.e., increased inclination of the frontal layer, can make it dynamically unstable. These several mechanisms argue that shear instability plays a major role in relaxing the frontal zone of coastal upwelling.

Part of the fascination of the coastal upwelling front rests in the fact that shoreward of the front there is relatively dense, recently upwelled water, whereas seaward of the front there is relatively light surface water. Yet, offshore Ekman transport must continue unless the winds reverse. It follows then that the offshore transport of the nearshore, upwelling water must be either transformed rapidly by solar heating, etc. in order to cross the front without sinking or there must be convective overturning and sinking at the surface front. The other alternative is for the surface front to move seaward indefinitely, but it does not.

Biological significance. Part of the importance of the coastal upwelling front lies in the fact that nutrient rich waters are brought into the euphotic zone inshore the front. But inshore of the front, the vertical (static) stability is much weaker than offshore of the front, thus the effective light levels are lower (through vertical stirring, etc.) inshore than offshore of the front. Apparently because substantial amounts of nutrients mix across the front into a shallow mixed layer zone which is bounded below by a strong pycnocline, the most intense concentrations of phytoplankton, *ergo* the most intense primary productivity, occur just seaward of the front. Associated with the phytoplankton, there are large concentrations of zooplankton, fish, marine mammals, and birds in the vicinity of the front. Benthic communities also flourish beneath the frontal zone due to the rich detrital food supply. To the extent that the frontal secondary circulation provides a recirculation mechanism for plankton detritus, nutrients, etc., these materials may be trapped in the coastal upwelling system, which may help explain the extraordinary biological productivity of such regions.

Circulation during upwelling and energy sources. Seasonal coastal upwelling is setup rapidly (in a few days). Once the permanent pycnocline has been elevated over the continental shelf, its main features only vary slowly over the upwelling season. However, in the nearshore upwelling zone, the pycnocline "flops around" substantially over wind event cycles (ca. 10 days in duration). Thus, while seasonal coastal upwelling elicits a primarily baroclinic response, the response to transient coastal upwelling is largely barotropic. As a consequence of the seasonal baroclinicity, an intense nearsurface, equatorward coastal jet is formed with its core located just offshore of the surface front. As a "complement" to the nearsurface equatorward jet, a seasonal averaged poleward undercurrent occurs nearbottom over the mid-shelf to upper slope regions.

The undercurrent provides another recirculation pathway for passive substances; e.g., phytoplankton, and for swimmers (fish and zooplankton) to use in their survival strategy. The transient, quasi-barotropic alongshore flow induced by transient coastal upwelling can be so strong as to overwhelm the equatorward surface jet and undercurrent so that the entire water column can either flow poleward or equatorward at some instants. The circulation response to upwelling wind events is complex.

A portion of the response is local and two-dimensional (mass balance in the cross-shelf vertical plane). Another portion can be large scale and represents the resonant response of coastally trapped long waves to atmospheric disturbances with favorable space-time scales. Yet another portion can be due to a forced, non-resonant wave response due to propagating atmospheric disturbances with alongshore phase velocities not favorable for resonance.

The coastally-trapped waves are quasi-barotropic over the shelf but quite baroclinic (bottom trapped) over the upper slope. They have alongshore coherence scales of the order of an alongshore wavelength (ca. 100 to 1,000 km) or less. Thus, they may actually be thought of as eddies, or eddy-waves. In some complex way, the transient upwelling response of coastally-trapped waves (free and forced) may be one of the long sought sources of turbulent eddy energy in coastal upwelling regions. They also provide a deformation field which may be important for frontogenesis.

Another major source of turbulent energy appears to be tidal energy, especially internal tidal motions, and inertial-internal wave energy, especially near-inertial motions. There is mounting evidence for internal tidal motion being generated near the shelf break and propagating onshore where it contributes to shear instability, especially, in the frontal zone, and possibly to driving the mean circulation. Similarly, there is strong evidence for a portion of the inertial motion generated in the surface layer during transient upwelling propagating downward (into the interior) and offshore, producing shear instability in the frontal zone and contributing to stirring of the bottom layer. The energy density and energy flux of both the internal tidal and near-inertial motion are observed to wax and wane over a wind event cycle.

A circulation model. As noted above, a conceptual model of shear instability, and thus mixing is emerging for the frontal zone. This means that a physical basis for specifying a space-time variable eddy coefficient (function) may arise from first principles. Already evidence exists for the inclined pycnocline to be interpreted as a frontal layer; i.e., an internal (turbulent) boundary layer which extends downwards and offshore from the surface layer, and whose turbulent intensity decays offshore. This internal boundary layer plays a role in driving cross-shelf circulation, which may be crucial for the coastal upwelling ecosystem as a feedback mechanism. A hypothetical example is shown in Figs. 6 through 9 for a simple conceptual model. Due to an equatorward wind stress and offshore surface Ekman transport, the density field (Fig. 6) shows (shoreward)

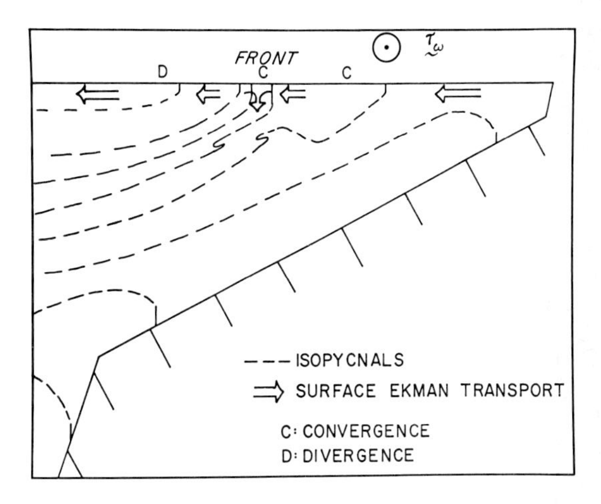

Fig. 6. Schematic diagram of density field and surface Ekman transport (in northern hemisphere) during active upwelling, viewed to the north. The arrowhead indicates equatorward wind stress.

elevated isopycnals; where also convergent, the isopycnals form a surface front at midshelf. The surface mixed layer is generally 5 to 10 m deep or less and quite variable in time, with a relatively deep zone just inshore of the surface front. The bottom mixed layer is generally 10 to 20 m thick or less and also quite variable in time. Convergent and divergent zones are shown in the surface layer; they are due to the effect of the front and associated jet in altering the Ekman transport. Some overturnings (density inversions) are shown below the surface front.

Corresponding qualitatively to the

density field, the alongshore (geostrophic) velocity (Fig. 7) is shown with an

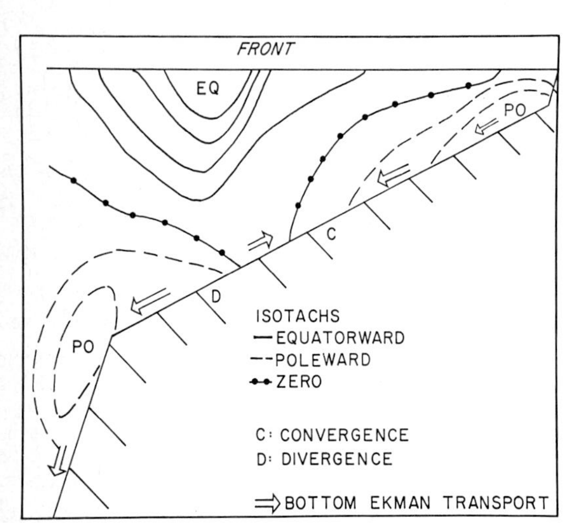

Fig. 7. Alongshore velocity field (in either hemisphere) during upwelling, viewed to the north.

equatorward surface jet and poleward undercurrents over the upper slope and outer shelf and nearshore. The surface jet has a strong cyclonic shear zone near the surface front and relatively strong vertical shear in the frontal zone. The bottom Ekman transport induced by the nearbottom alongshore flow produces covergent and divergent zones.

Internal tidal energy is generated near the shelfbreak and propagates onto the shelf (Fig. 8) where it contributes to bottom-stirring, is affected by "shoaling" in the frontal zone, and provides vertical shear for mixing in the frontal zone. Near-inertial motion, generated nearshore by wind transients, propagates downward and offshore along the frontal zone, providing further vertical shear for mixing. Altogether then, the frontal zone is seen to be a region of relatively high shear instability and, thus, turbulence, whose intensity decays offshore. Hence, a localized internal boundary layer is formed.

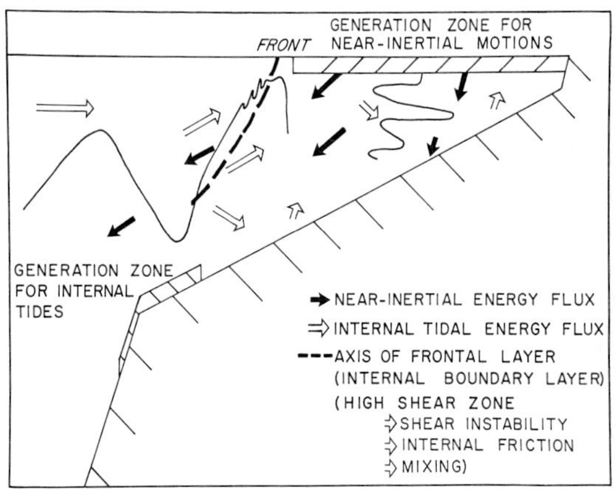

Fig. 8. Energy generation and flow in upwelling zone. The wavy line offshore represents a typical isopycnal; the wiggly line under the surface generation zone represents a downward propagating near inertial wave with high vertical shear in the horizontal velocity.

A cross-shelf circulation (Fig. 9) is generated by the combined effects of the Ekman transport in the surface, bottom, and internal boundary layers. Since the alongshore velocity has an extremely variable quasi-barotropic component, due to the variable wind stress, the Ekman transports are variable and thus, this circulation pattern is only a "snapshot" of the time variable cross-shelf circulation system. However hypothetical this analysis may be, it is suggestive of the diagnostic possibilities in frontal zones where geostrophic flows and Ekman transports are significant.

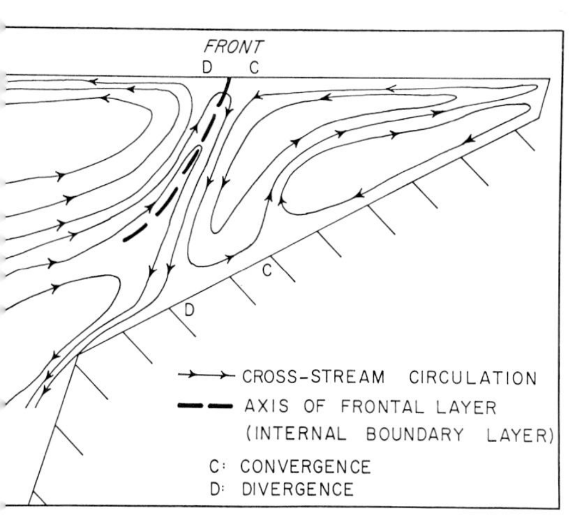

Fig. 9. Cross-shelf circulation (in either hemisphere) during upwelling, viewed to the north.

Future directions. The greatest weakness in coastal upwelling field studies to date is the failure to resolve the frontal scale in a three-dimensional, time series fashion. The required domain is of the order of 10 km : cross-shelf, 100 km : alongshelf, 100 m : vertical; and 10 da. : temporal. The required resolution is of the order of 1 km : cross-shelf, 10 km : alongshelf, 5 m : vertical, and 1 hr : temporal. The resolution requirements may be applied non-uniformly and, thus, the sampling requirements made practical. Also, Lagrangian as well as Eulerian tools may be used for optimization of the information-to-effort ratio.

Another weakness in coastal upwelling studies is the lack of a circulation model which relates the various physical fields in a dynamically consistent fashion, including the frontal zone.

The preceeding discussion of coastal upwelling fronts is partially based upon material in the papers listed in the references.

References

Retrograde Fronts

Beardsley, R. C. and B. Butman. 1974. Circulation on the New England continental shelf: response to strong winter storms. Geophys. Res. Let., 1, (4) 181-184.

Beardsley, R. C. and C. N. Flagg. 1975. The water structure, mean currents, and shelf-water/slope-water front of the New England continental shelf. Seventh Liege Colloquium on Ocean Hydrodynamics, 10, 209-226.

Beardsley, R. C., W. C. Boicourt, and D. V. Hansen. 1976. Physical oceanography of the Middle Atlantic Bight. J. Limnol. Oceanogr., Spec. Symp., 2, 20-34.

Beardsley, R. C., H Mofjeld, M. Wimbush, C. Flagg, J. Vermersch, Jr. 1977. Ocean tides and weather induced bottom pressure fluctuations in the Middle Atlantic Bight. J. Geophys. Res., 82 (21), 3175-3182.

Bigelow, H. B. 1933. Studies of the waters on the continental shelf, Cape Cod to Chesapeake Bay I, the cycle of temperature. Massachusetts, Institute of Technology and Woods Hole Oceanographic Institution Papers in Phys. Oceanog. and Meteor., 2, 4.

Bigelow, H. B. and M. Sears. 1935. Studies of the waters on the continental shelf, Cape Cod to Chesapeake Bay II, Salinity. Massachusetts Institute of Technology and Woods Hole Oceanographic Institution Papers in Phys. Oceanog. and Meteor., 4, 1.

Boicourt, W. C. 1973. The circulation of water on the continental shelf from Chesapeake Bay to Cape Hatteras. Doctoral Dissertation, John Hopkins University.

Boicourt, W. C. and D. W. Hacker. 1976. Circulation on the Atlantic continental shelf of the United States, Cape May to Cape Hatteras. Memoires de la Societe des Sciences de Liege, 16 series, 10, 187-200.

Bowman, M. J. and L. D. Wunderlich. 1977. Hydrographic properties. MESA New York Bight Atlas monograph 1. New York State Sea Grant Institute, Albany, N.Y.

Bumpus, D. F. 1973. A description of the circulation on the continental shelf of the east coast of the United States. Prog. in Oceanog., 6, 111-158.

Butman, B. 1975. On the dynamics of shallow water currents in Massachusetts Bay and on the New England continental shelf. Doctoral Dissertation.

Chamberlin, J. 1976. Bottom temperatures

on the continental shelf and slope south of New England during 1974. Atlantic Environmental Group Preprint.

Cresswell, G. M. 1958. The quasi-synoptic monthly hydrography of the transition region between coastal and slope water, to the south of Cape Cod, Massachusetts. Unpublished Manuscript, Woods Hole Oceanographic Institution Reference No. 67-35.

Flagg, C. N. and R. C. Beardsley. 1975. 1974 M.I.T. New England shelf dynamics experiment (March, 1974). Part I: hydrography. M.I.T. report 75-1.

Flagg, C. N. 1977. The kinematics and dynamics of the New England continental shelf and shelf/slope front. Doctoral Dissertation MIT/WHOI.

Ingham, M. C. 1976. Variations in the shelf water front off the Atlantic Coast between Cape Hatteras and Georges Bank. U.S. Dept. of Commerce, MARMAP Contribution No. 140 (unpublished).

Morgan, C. W. and J. M. Bishop. 1976. Eddy inducted water exchange along the continental slope. J. Phys. Oceanog.

Saunders, P. 1971. Anticyclonic eddies formed from shoreward meanders of the Gulf Stream. Deep Sea Res., 18, 1207-1219.

Voorhis, A. D., D. C. Webb and R. C. Millard. 1976. Current structure and mixing in the shelf/slope water front south of New England. J. Geophys. Res., 81 (21) 3695-3708.

Wright, W. R. 1976. The limits of shelf water south of Cape Cod. J. Mar. Res., 34, 1, 1-14.

Prograde Fronts

Curtin, T. B. and C. N. K. Mooers. 1975. Coastal upwelling experiments I and II, Surface Hydrographic Fields Data Report. RSMAS, University of Miami, Miami, FL., Ref. No. 74026, 94 pp.

Curtin, T. B. and C. N. K. Mooers. 1975. Observation and interpretation of a high-frequency internal wave packet and surface slick pattern. J. Geophys. Res., 80 (6) 882-894.

Curtin, T. B., W. R. Johnson and C. N. K. Mooers. 1975. Coastal upwelling experiment II. Hydrographic Data Report. RSMAS, University of Miami, Miami, FL. Ref. No. 75003, 100 pp.

Hayes, S. P. and D. Halpern. 1976. Observations of internal waves and coastal upwelling off the Oregon Coast. J. Mar. Res., 34, 247-267.

Johnson, D. R., E. D. Barton, P. Hughes, and C. N. K. Mooers. 1975. Circulation in the Canary Current upwelling region off Cabo Bojador in August, 1972. Deep Sea Res., 22 (8) 547-558.

Johnson, D. R. 1977. On double-cell circulation during coastal upwelling (in preparation).

Johnson, W. R., J. C. Van Leer, and C. N. K. Mooers. 1976. A cyclesonde view of coastal upwelling. J. Phys. Oceanogr., 6, 556-574.

Johnson, W. R. and C. N. K. Mooers. 1977. A case study of inertial-internal waves in a coastal upwelling region. J. Phys. Oceanogr. (to be submitted).

Kundu, P. 1976. An analysis of inertial oscillations observed near the Oregon Coast. J. Phys. Oceanogr., 6, 879-893.

Mooers, C. N. K. 1975. Several effects of a baroclinic current on the cross-stream propagation of inertial-internal waves. Geophys. Fluid Dyn., 6, 245-275.

Mooers, C. N. K., C. A. Collins, and R. L. Smith. 1976. The dynamic structure of the frontal zone in the coastal upwelling region off Oregon. J. Phys. Oceanog. 6, 3-21.

Stevenson, M. R., R. W. Garvine, and B. Wyatt. 1974. Lagrangian measurements in a coastal upwelling zone off Oregon. J. Phys. Oceanogr., 4 (3) 321-336.

Wang, D.-P. and C. N. K. Mooers. 1976. Coastal-trapped waves in a continuously stratified ocean. J. Phys. Oceanogr., 6 (6) 853-863.

Wang, D.-P. and C. N. K. Mooers. 1977. Long coastal-trapped waves off the west coast of the United States, Summer 1973. (Resubmitted) J. Phys. Oceanogr.

Wang, D.-P. and C. N. K. Mooers. 1977. Evidence for interior dissipation and mixing during a coastal upwelling event off Oregon. J. Mar. Res., 35 (4) (to appear Nov. 1977).

7. PHYSICAL ASPECTS OF THE NOVA SCOTIAN SHELFBREAK FRONTS

EDWARD P. W. HORNE

Introduction

The existence of a boundary zone within the subsurface slope water off Nova Scotia was first suggested by Fuglister (1963) after analyzing "Gulf Stream '60" data. This topic was investigated further by Gatien (1976) who upon closer examination of the "Gulf Stream '60" data concluded that indeed there are two distinct water masses, which she called warm slope water and Labrador slope water, separated by a sharp boundary (Fig. 1). Exchange of water properties across this front is clearly important to any consideration of overall budgets for mass transfer in this area, and it was suggested that processes at the front might give rise to a vertical transport of nutrients to help account for the high productivity in the warm surface slope water (Fournier *et al.*, 1977). In this section, I describe results obtained from a detailed hydrographic survey of this boundary zone and the physical interpretation of the data. The location of the hydrographic survey is illustrated in Figure 2. Further details can be found in Horne (in press).

Hydrographic features

In all previous sections taken across

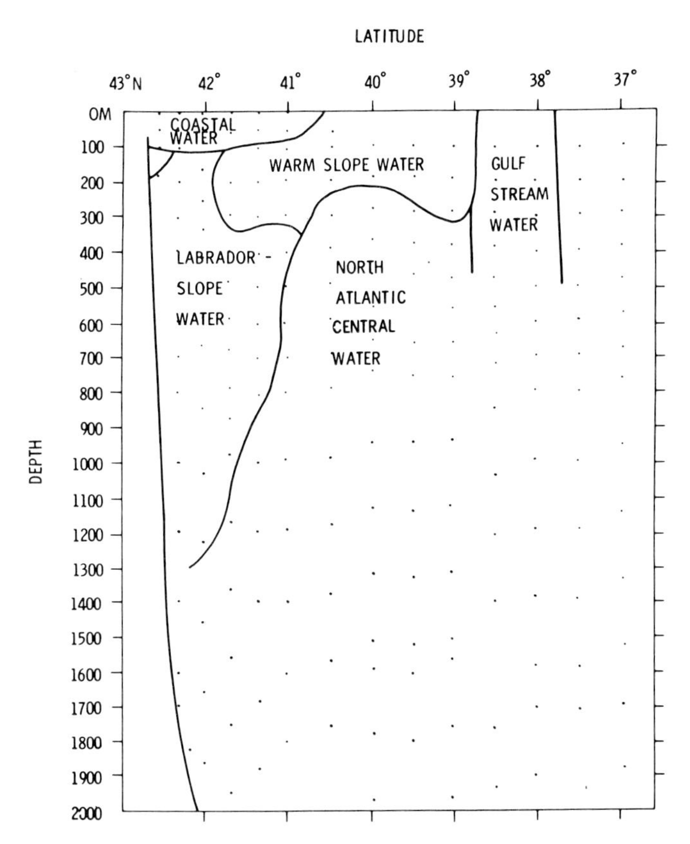

Fig. 1. Approximate positions of the two Slope Water zones for Section III of "Gulf Stream '60" (from Gatien 1976).

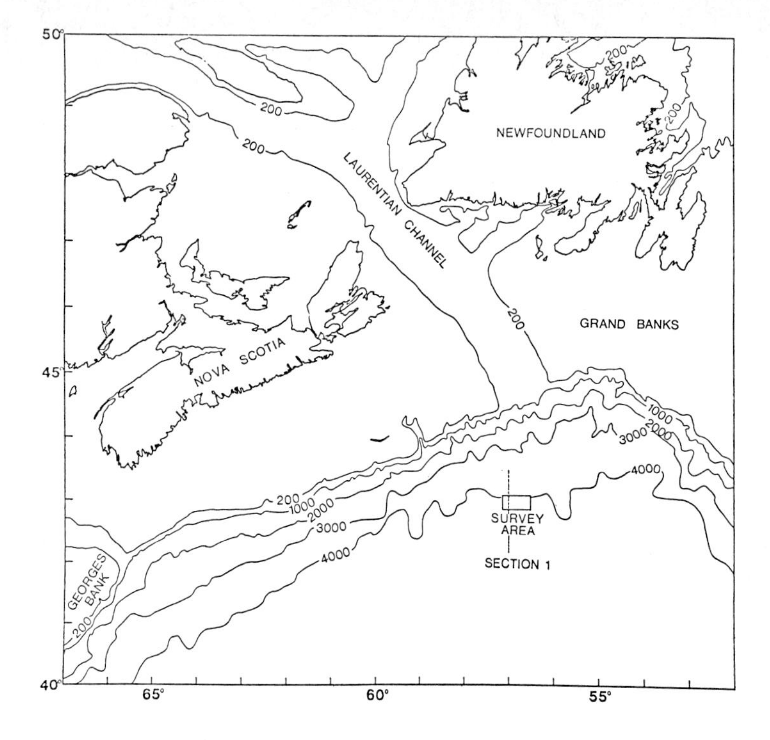

Fig. 2. Chart showing survey area.

this boundary zone the station spacing was never less than about 10 km. Figures 3 and 4 show vertical sections of temperature and sigma-t taken in April 1975. A layer of warm slope water bounded approximately by the 7°C isotherm intrudes shoreward underneath the coastal water (which is confined to the top 75 m and reaches its seaward extremity at about st. 13). The boundary zone at st. 7 is most noticeable in the temperature section at the 10°C isotherm where the width of the *subsurface* front is less than 10 km. The density change across this front is small (Fig. 4) since temperature and salinity (not shown) are largely compensatory.

The front is a permanent feature of the area; its onshore-offshore location can vary by at least 150 km over a few months. To describe the structure of the front in detail, the ship was allowed to drift across the front with the CTD continuously cycled up and down as quickly as possible.

The predominant features in the profiles across the frontal zone are the strong interleaving of the water masses with a vertical scale of 10-20 m (Figs. 5-7). Down traces only are shown; the up traces are noisy since the sensors are in the wake of the instrument. Each profile took about 10 minutes, during which the ship drifted less than 500m.

Notice the feature labeled A (Figs. 5-6) that first appears in profile 14 at a depth ~ 100 dbar. There are changes in temperature of 1.0°C and salinity of 0.2‰ over a horizontal distance of less than 500m. Notice also that individual layers remain coherent from one profile to the next for several kilometers. The

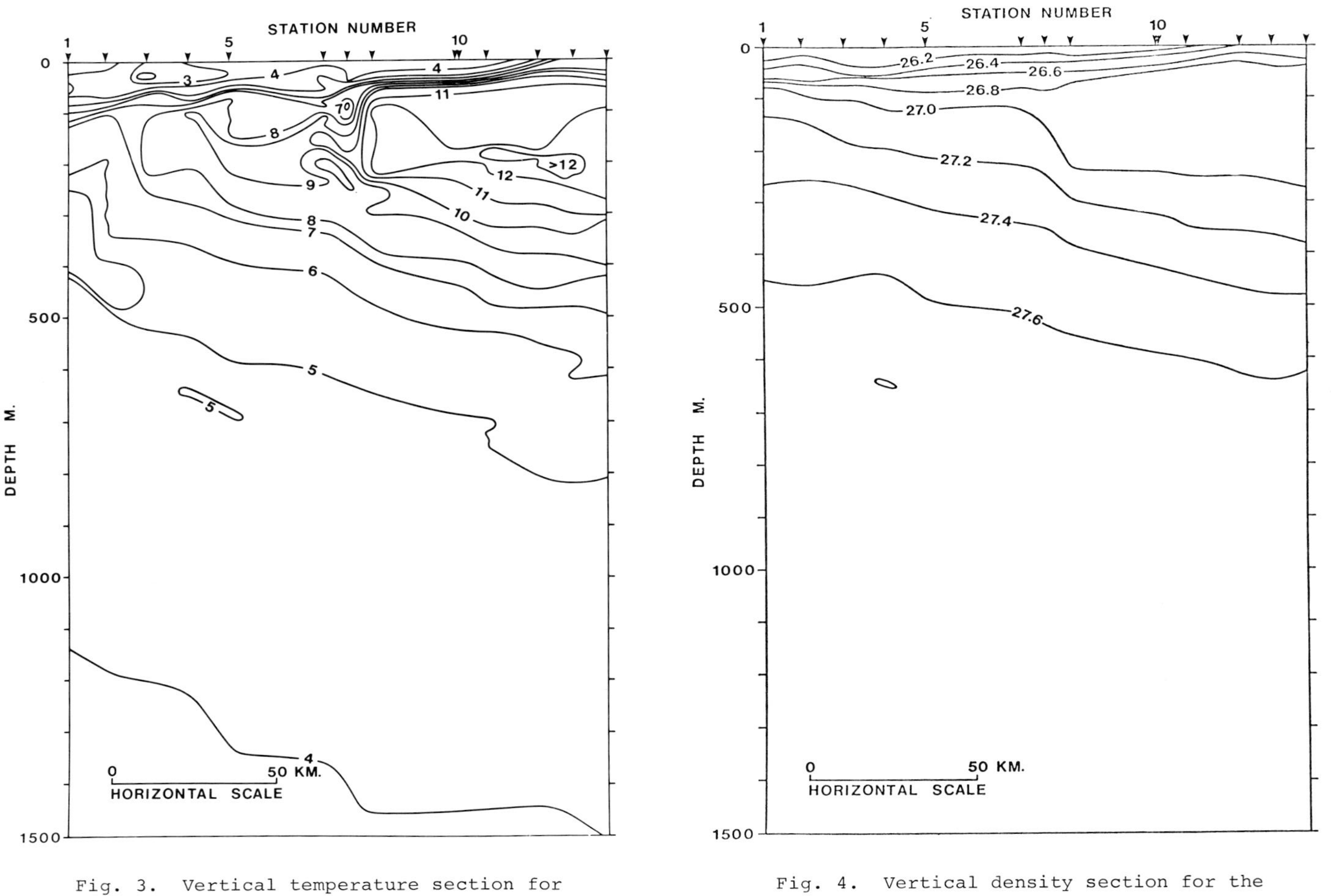

Fig. 3. Vertical temperature section for the line marked section 1 on Fig. 2.

Fig. 4. Vertical density section for the line marked section 1 on Fig. 2.

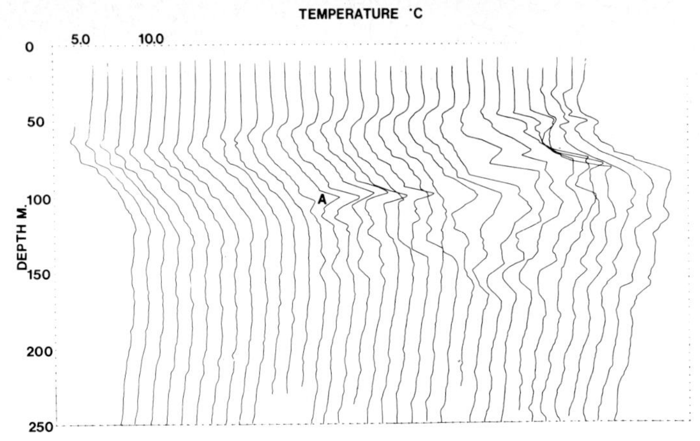

Fig. 5. Temperature profiles for Station 62. The horizontal scale is correct for the first profile and each additional profile is offset 1°C.

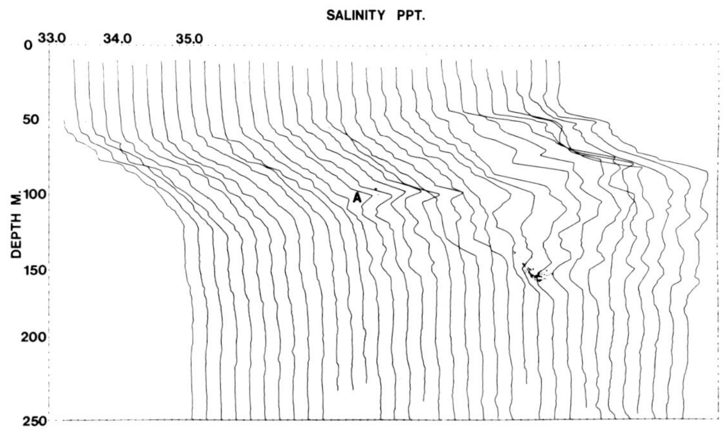

Fig. 6. Salinity profiles for Station 62. The horizontal scale is correct for the first profile and each additional profile is offset 0.2°/oo.

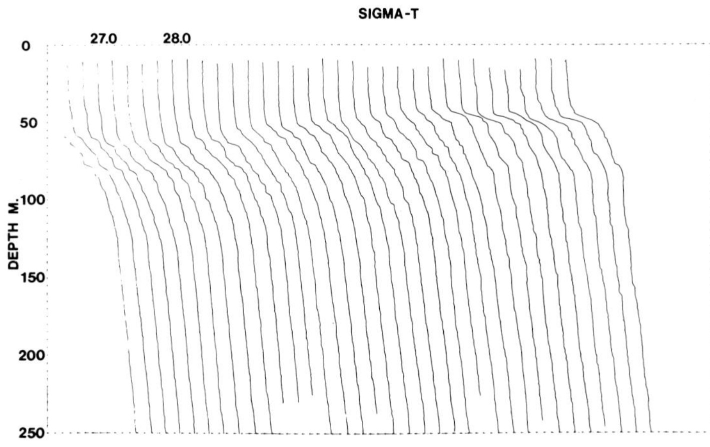

Fig. 7. Density profiles for Station 62. The horizontal scale is correct for the first profile and each additional profile is offset 0.2 σ_T.

density profiles, in contrast, show no comparable gradients between profile numbers 13 and 14. However, they do exhibit a step-like structure thus showing that the layers are homogeneous. The salinity distribution across the frontal zone (Fig. 8) has a compensating structure to the temperature field (not shown), such that almost no structure is present in the density (Fig. 9).

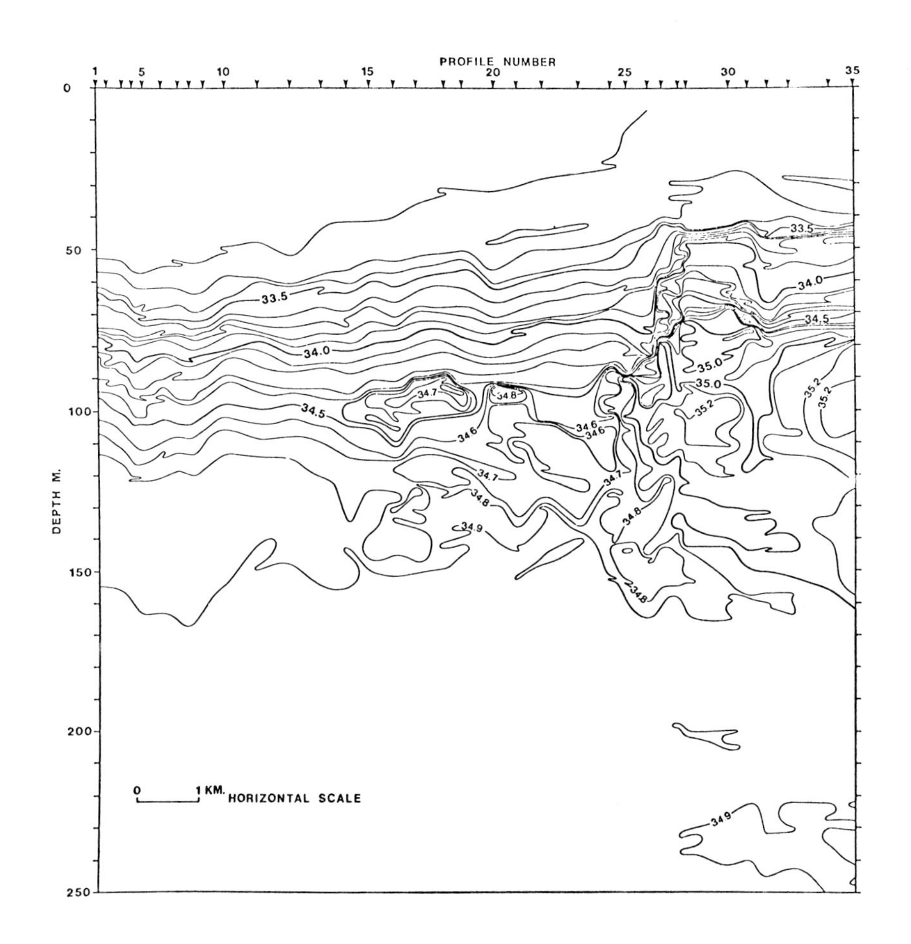

Fig. 8. Vertical salinity sections for Station 62.

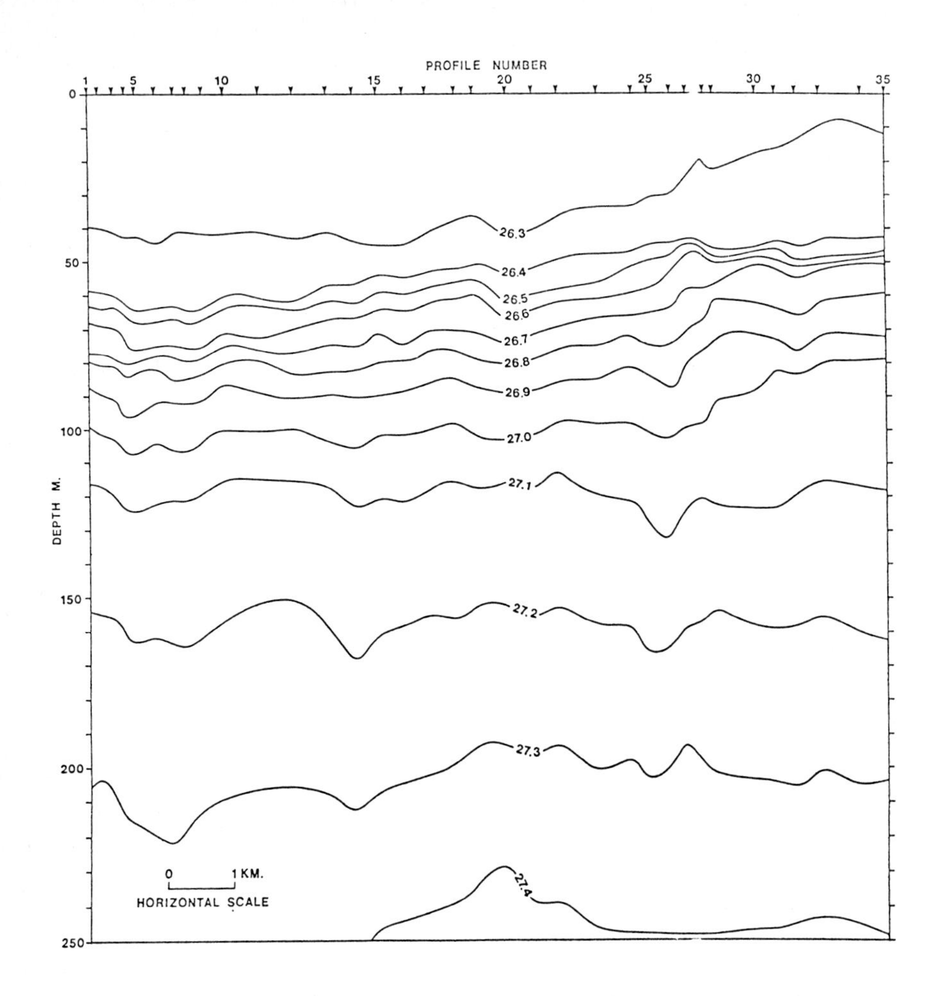

Fig. 9. Density sections for Station 62.

Lifetime and velocity of interleaving layers

An estimate of the lifetime of these interleaving layers can be obtained by considering a vertical column of water of height 2h and unit cross-section which is losing heat and salt through its ends by double diffusive fluxes as it moves along with velocity v (Fig. 10). If we take typical experimental values for the initial temperature and salinity differences (ΔT_0 and ΔS_0) between the center of the layer and outside of both interfaces then fluxes of heat and salt through the ends of the column can be calculated (see section 11).

The lifetime of the layer can be approximated as the time required for these fluxes to reduce the temperature anomaly to zero. With $\Delta T_0 \sim 2.1°C$ and $\Delta S_0 \sim .5°/_{\circ\circ}$ (layer A; Figs. 5 and 6) the lifetime of this layer τ is ~ 31 hours. From Fig. 5 the cross-frontal extent of layer A is ~ 3 km which implies a

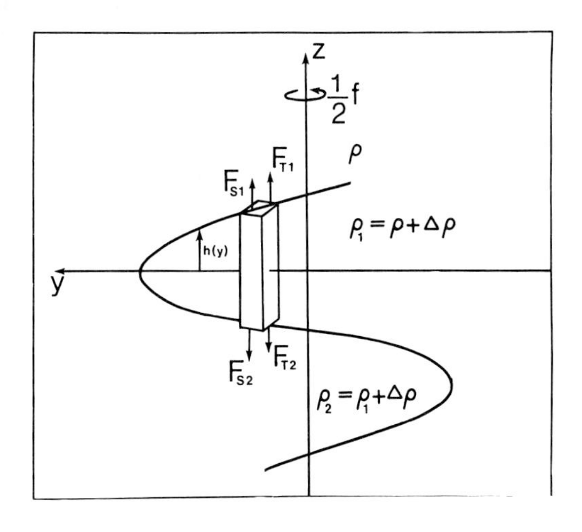

Fig. 10. A typical interleaving layer with heat and salt fluxes (F_S and F_T) through its upper and lower interfaces. h(y) is the half height of the layer, ρ_1 is the density of the warm slope water and ρ_2 the density of the coastal water. $\frac{1}{2}f$ represents the earth's rotation.

velocity, v, of ~ 3 cm sec^{-1}.

Little research has been undertaken on the dynamics of intrusive features. Stommel and Fedorov (1967) demonstrated the importance of friction in limiting the horizontal spread of layers. They showed that, for small Ekman numbers, Ekman layers form at the interfaces of the layers while for large Ekman numbers, the motion of the layers is governed by viscous dynamics. In both cases, the rate of change of layer thickness can be approximately described by a diffusion equation. The lateral transport, and hence the effective diffusivity, tends to zero at both large and small Ekman numbers, because for small Ekman numbers the layers are too thin to have significant transport, while for large Ekman numbers the fluid is too viscous to move at significant velocities.

Using Stommel and Fedorov's method of estimating horizontal diffusivity K_H (see section 11), a value for intrusion A of ~ 2.6 x 10^5 cm^2 sec^{-1} is obtained.

The horizontal length scale associated with this value of K_H is given by $(K_H\tau)^{\frac{1}{2}}$ ~ 2 km. This is in good agreement with observations illustrated in Figs. 5-6.

<u>Circulation in the frontal zone</u>

Velocity measurements were taken across the front using window blind drogues set at 80 m. The trajectories are shown in Figure 11. The position of the front was determined from XBT profiles, and was found to lie between drogues 2 and 3.

The results were somewhat inconclusive as during the first half of the tracking there did not appear to be any shear or convergence across the front, while during the second half of the tracking there may have been some shear across the front (2 cm sec^{-1}) as well as some convergence. The mean current measured along the front was ~ 20 cm sec^{-1}. Taking the layer of no motion at 500 m, the geostrophic shear across the front was calculated to be ~ 9 cm sec^{-1}. The reason why the measured shear was not as great as that calculated by geostrophy is probably because the drogues were not situated within the main density gradient (~ 200 m) (Fig. 4).

Cabbeling in the frontal zone is a prime candidate as a driving mechanism (section 11). It is suggested that the nutrient flux required to sustain the high biomass at the surface front can be driven by the high vertical eddy diffusivity in the frontal zone (~ 5-50 $cm^2 sec^{-1}$), in spite of downwelling induced by cabbeling.

<u>Along frontal coherence</u>

The along-frontal coherence of the layers can be estimated by examining Figure 12 which shows T-S curves for three profiles spaced over a total distance of ~ 18 km along the front. Four prominent

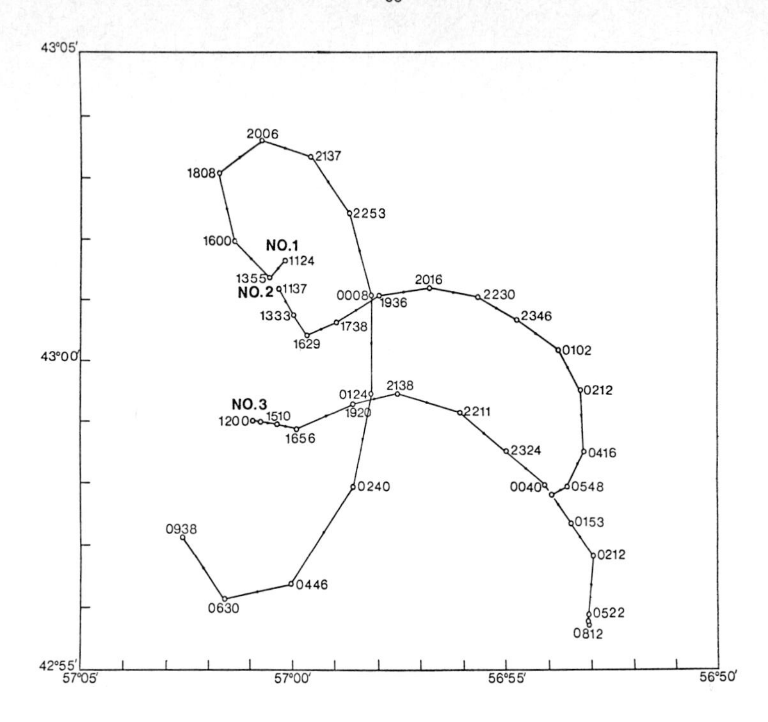

Fig. 11. Drogue paths at 80 m. The front is between drogue 2 and 3. Drogue 1 broke off from its surface penetrator at some time during the experiment so its data cannot be interpreted.

interleaving layers are coherent over this distance. This suggests that the layers originate as sheets, and not as tongues, which are subsequently stretched in the along-front direction by velocity shear as suggested by Voorhis *et al.* (1976).

Contributions to heat and salt budgets

Nova Scotian shelf. An estimate of the importance of mixing across this front to the Nova Scotian shelf waters can be obtained by repeating the calculations of Voorhis *et al.* (1976) for the coastal water slope water front along the New England Shelf. Using a value of the mean vertical diffusivity for salt $K_H(S)$ ~ 3 cm^2 sec^{-1} and mean vertical salinity gradient $\partial S/\partial z \sim 1.5 \times 10^{-4}$ ‰ cm^{-1} we obtain a vertical salt flux $K_H(S) \cdot \partial S/\partial z \sim 4 \times 10^{-2}$ gm cm^{-2} day^{-1} (0.4 kg m^{-2} day^{-1}).

Taking the frontal width to be 10 km and the length of the Scotian Shelf to be 970 km, then this flux will add 1.4×10^{12} kg of salt to the coastal water each year. The major source of fresh water for the Scotian Shelf is the St. Lawrence River which has an annual discharge of ~ 400 km^3/yr (Trites, 1970). The salt flux calculated above is enough to raise its salinity from 0 to 3.2 ‰.

The same calculation for heat shows that 10×10^{18} cal are added to the coastal water annually. The total monthly solar radiation for Halifax was obtained from the Monthly Record (a publication of meteorological observations in Canada put out by Environment Canada) to compare this with the heat input from insolation. The average monthly albedo at the sea surface

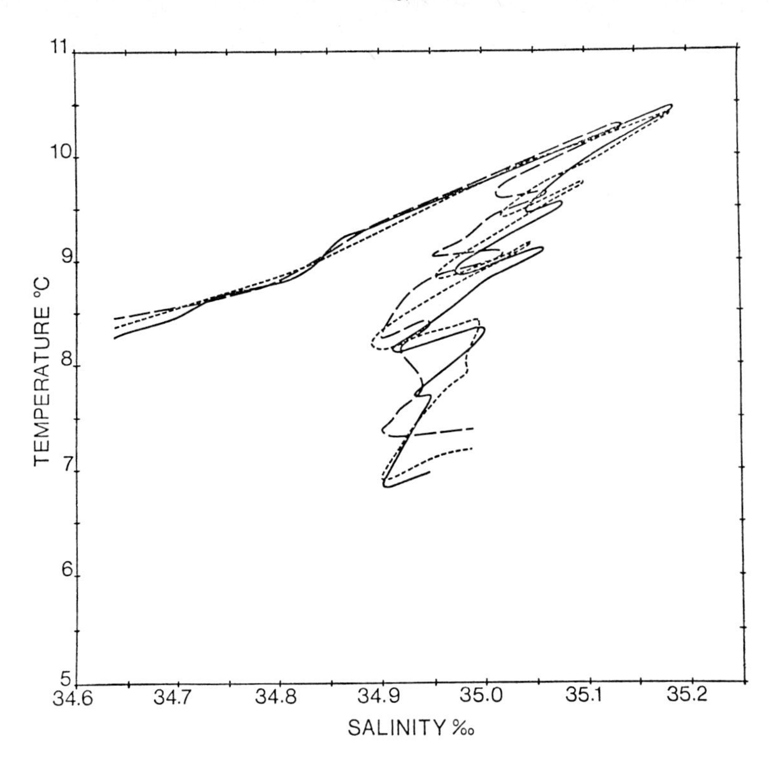

Fig. 12. T-S curves for profile 12 of Station 58 (short dashes), profile 33 of Station 62 (solid), and profile 9 of Station 63 (long dashes), showing along-frontal coherence.

was obtained from Payne (1972). Estimating the area of the Scotian Shelf as $\sim 2 \times 10^5$ km^2 leads to an annual solar input of 2×10^{20} cal. Since the frontal heat flux is some 15% of this and the salt flux can account for about 10% of the salt content of the coastal water, exchange through the shelfbreak frontal zone is a major contributor to the heat and salt budgets of the shelf waters.

Taking the thickness of the front to be 200 m, $\sim 1.5 \times 10^{11}$ cal and 3.0×10^4 kg salt are added daily to a column of LSW 1m wide. The average westward velocity of the Labrador slope water (LSW) is about 3 cm sec^{-1} (Petrie, personal communication). It will therefore take a parcel of LSW $\sim$ 400 days to travel the length of the Scotian shelf, during which time it will gain $\sim 6 \times 10^{13}$ cal and $\sim 1 \times 10^7$ kg salt. Taking the cross sectional area of the LSW to be 2.5×10^8 m^2, its average temperature to be 0.5°C cooler, and its average salinity to be 0.1°/$_{oo}$ less than the slope water, we find that the above fluxes can account for $\sim$ 45% of the heat and $\sim$ 40% of the salt required to modify the LSW during the time it takes to traverse the Scotian shelf. These calculations indicate that we can account for the fact that LSW does not appear farther west than Georges Bank.

Acknowledgements

Thanks are expressed to C. J. R. Garrett, G. T. Needler, P. C. Smith, the entire Ocean Circulation Division of the Bedford Institute of Oceanography, and numerous people from Dalhousie University.

This project was supported by the National Research Council of Canada.

References

Fournier, R. O., J. Marra, R. Bohrer, and M. Van Det. 1977. Plankton Dynamics and Nutrient Enrichment of the Scotian Shelf. J. Fish. Res. Board Can. 34(7):1004-1018.

Fuglister, F. C. 1963. Gulf Stream '60. Prog. Oceanogr. 1:265-373.

Gatien, M. G. 1976. A Study in the Slope Water Region South of Halifax. J. Fish. Res. Board Can. 33(10):2213-2217.

Horne, E. P. W. in press. Interleaving at the subsurface front in the slope water off Nova Scotia. (to be published in J. Geophys. Res.).

Payne, R. E. 1972. Albedo of the Sea Surface. J. of the Atm. Sci. 29: 959-970.

Stommel, H., K. N. Fedorov. 1967. Small scale structure in the temperature and salinity near Timor and Mindanae. Tellus 19:306-325.

Trites, R. W. 1970. The Gulf of St. Lawrence from a pollution viewpoint. FAO technical conference on marine pollution and its effects on living resources and fishing. FIR; MP/70/ R-16.

Voorhis, A. D., D. C. Webb, and R. C. Millard. 1976. Current Structure and Mixing in the Shelf/Slope Water Front South of New England. J. Geophys. Res. 81(21):3695-3708.

8. BIOLOGICAL ASPECTS OF THE NOVA SCOTIAN SHELFBREAK FRONTS

ROBERT O. FOURNIER

Introduction

Although fronts have been recognized as oceanographic phenomena since before the turn of the century, oceanographers have devoted little effort toward understanding the biological significance of these boundary regions. Uda (1959), in addition to compiling a summary of fronts and frontal physical characteristics, also described associated biological phenomena observed in the coastal seas of Japan. He suggested that fronts act as faunal and floristic boundaries limiting the distribution of organisms in much the same way as "zoogeographical features of the shores and sea bottoms". He also reported that frontal convergences resulted in localized accumulations of plankton and particulate materials which in turn attracted higher levels in the food chain--a fact fully exploited by Japanese fishermen. Uda also suggested that, in addition to the accumulating effect of two convergent water masses, the fronts have a stimulatory effect on phytoplankton growth, particularly in the spring. These observations were qualitative but they strongly suggested that frontal biology should be an exciting and productive field of research.

This contribution is a brief summary of work done off the coast of Nova Scotia with an emphasis on late winter, early spring conditions (further details will be published in the proceedings of the Chapman Conference on oceanic fronts, New Orleans, October 11-13, 1977).

Characteristics

The shelfbreak front off Nova Scotia is located in the transition zone between continental shelf coastal water and denser, offshore slope water. Gradients in temperature and salinity across the front, often of complex structure, can vary by as much as 6°C and 2°/₀₀ over as little as 3-5 km. Although the exact physical location of the front can vary by tens of kilometers over a period of days to weeks its average position appears to be parallel to the edge of the continental shelf lying over the 100 to 1000 m isobaths.

The shelfbreak zone is a highly productive fishing area. Figures 1 and 2 show the distributions of Canadian and foreign fishing vessels as spotted by reconnaissance aircraft for one year (1972). These vessels are clearly concentrated along the shelf edge and lend credence to the frontal zone possessing highly significant biological properties.

Figure 3 shows the mean locations of major ocean currents off the east coast of Canada plus the standard station positions sampled by the Institute of Oceanography along a transect perpendicular to the coastline.

Sampling Cruises

Sampling cruises along these transects from Halifax into slope water have been made in all seasons. The data have shown a reasonably consistent pattern of high relative fluorescence associated with sharp thermal gradients which define the fronts (Fournier *et al.*, 1977). On two of these cruises, high relative fluorescence was observed at the front during periods of strong vertical stability (May and August), as well as on a third occasion (March), when intense vertical mixing was present. Fournier *et al.* (1977) suggested that during these periods of stability (i.e., from the spring flowering to the fall overturn), when nutrient limitation is most severe, high standing stocks of phytoplankton most likely occur at the front because of a net nutrient flux upward in that localized region. The mechanism could be slow and steady or intermittent but would result from the dynamic upwelling properties of the front, ultimately yielding enhanced phytoplankton growth.

The Nova Scotia shelfbreak front acting as a nutrient pump presents a paradox,

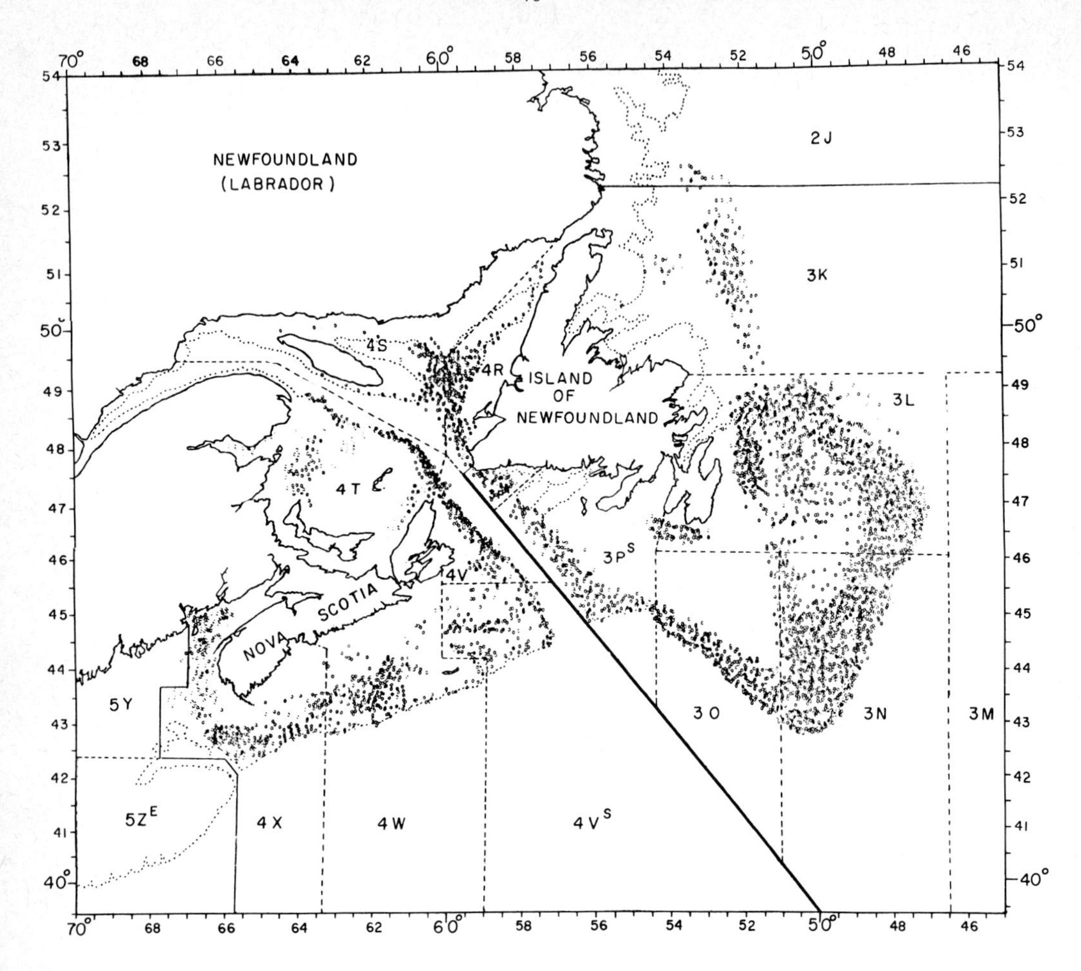

Figure 1. Representative distribution of Canadian fishing vessels for one year (1972) off the east coast of Canada (from Grant and Rygh, 1973).

since maximum relative biological activity occurs at the front in March when nutrients are in abundant supply in both shelf and slope waters. Enhanced vertical transport of nutrients at any specific location would not selectively enhance growth at that location since nutrients are not limiting.

In March 1977, a cruise was again taken along this transect into slope water (Figure 3) with the express intention of considering this paradoxical phenomena. The length of the transect was 370 km from Halifax Harbour to a point which usually is well into slope water. Figure 4 shows the outer 135 km of that transect beginning with the transition from coastal to slope water. Displayed in this figure are two replicate transects of continuous surface (1 m) temperature and relative fluorescence. The data presented in block "A" were obtained on a continuous run of approximately 24 hr. The data shown in block "B" were obtained on the return voyage. Seventeen stations at which hydrographic and biological observations were carried out were also made on this second run. Block "C" presents values for

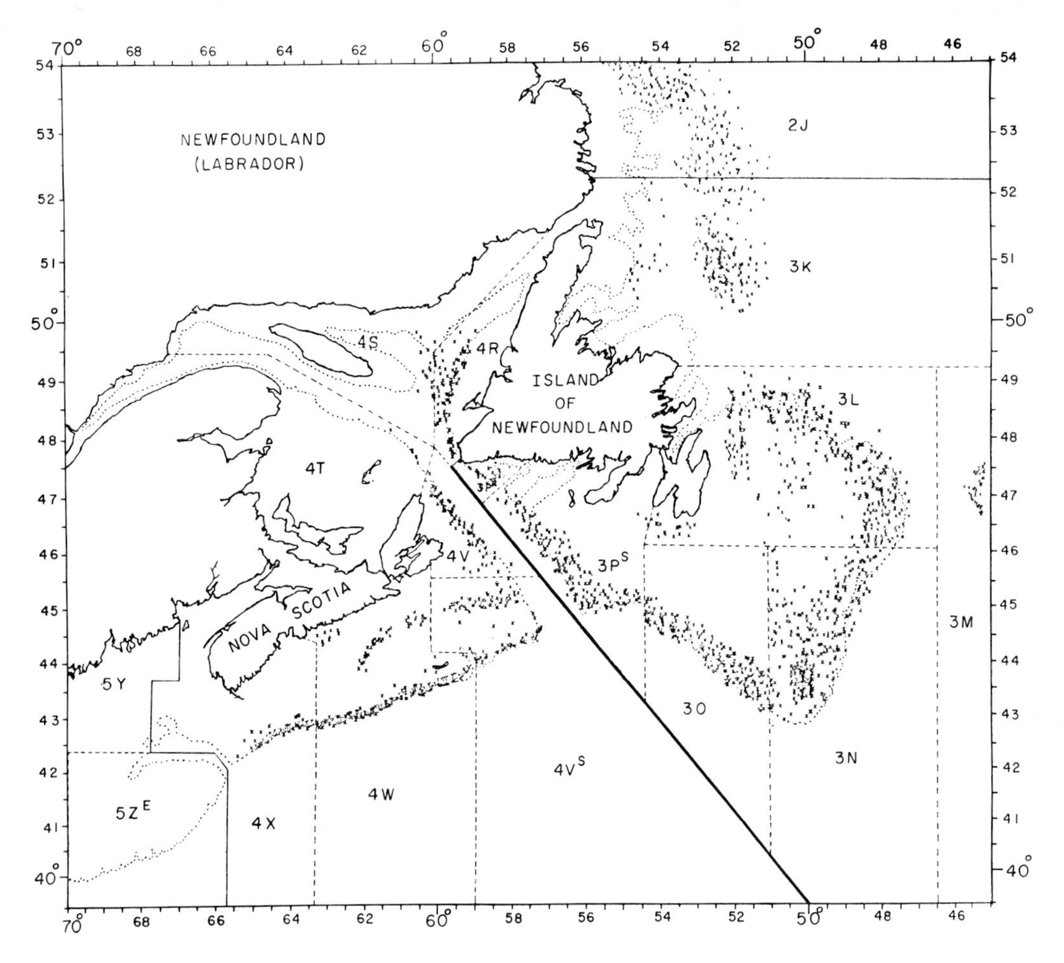

Figure 2. Representative distribution of foreign fishing vessels for one year (1972) off the east coast of Canada (from Grant and Rygh, 1973).

chlorophyll *a* extracted from water samples obtained at those stations.

The average value for chlorophyll *a* across the shelf was less than 1 mg m^{-3} whereas upon crossing the shelf/slope front (stations 16-21; temperature contrasts ~ 6.5°C) chlorophyll reached a maximum of ~ 4.2 mg m^{-3}. Chlorophyll dropped sharply at station 22, rose to intermediate values (along with temperature) between stations 25-28, then dropped again when temperatures rose to slightly more than 15°C across the outer front.

Discussion of results

In order to understand these changes it is necessary to consider the hydrography of this area. Figure 5 presents temperature, salinity and density sections for stations 16-33. These diagrams illustrate the hydrographic complexities of the region with multiple fronts, both prograde and retrograde (see Chapter 6), present.

The inshore retrograde front is clearly visible between stations 16-21 showing very sharp horizontal and vertical gradients in the upper 50 m for all three

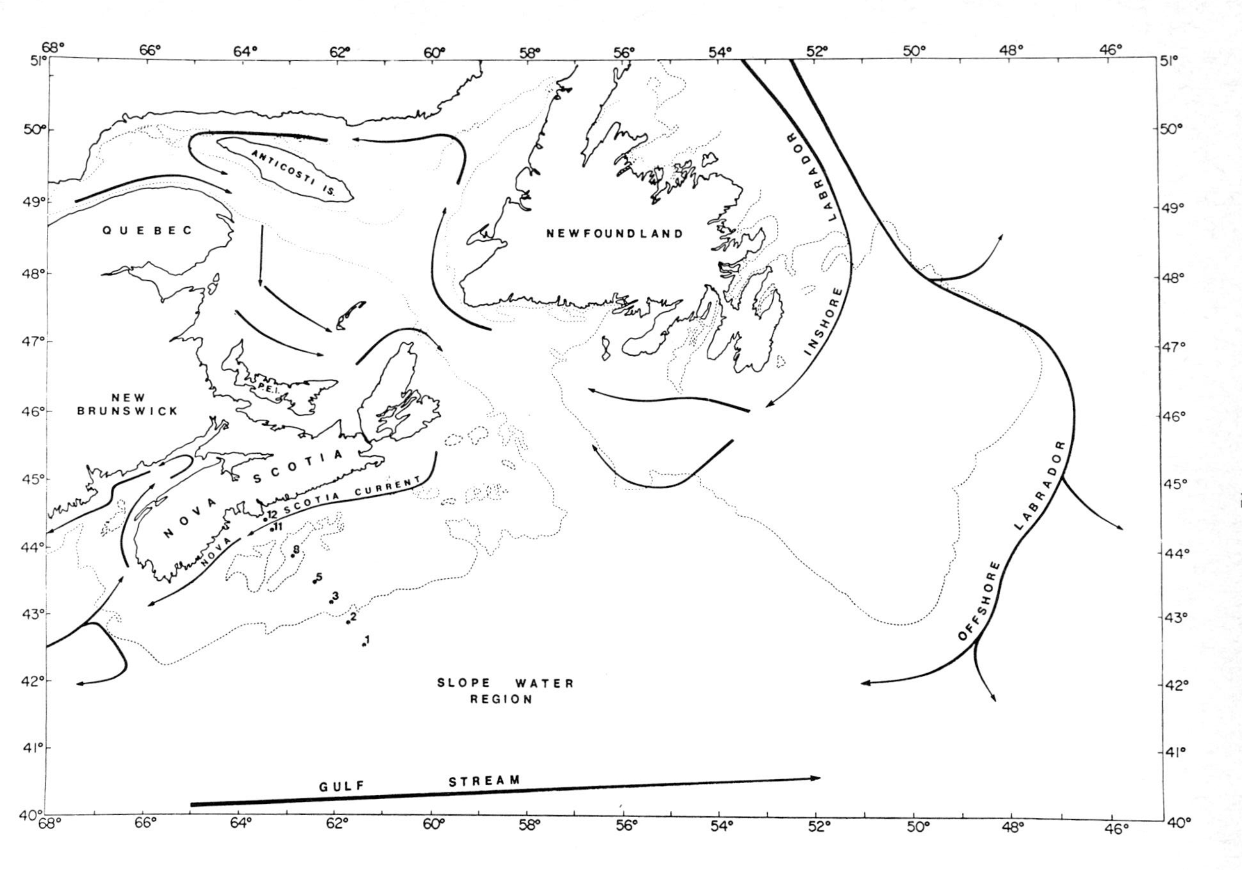

Figure 3. Major oceanic currents off the east coast of Canada, and the seven sampling stations across the Scotian shelf from Halifax Harbour. Dashed line is the 200 m contour.

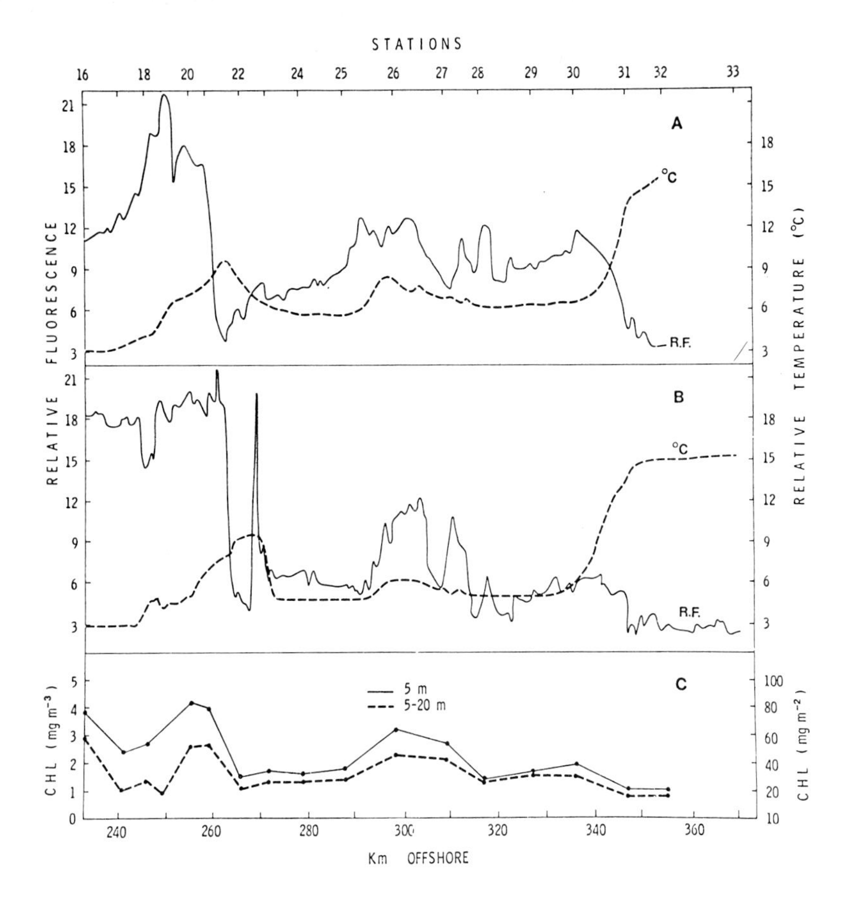

Figure 4. A. Continuous surface (1 m) temperature and relative fluorescence for outward run along transect, March 9-10, 1977.

B. Continuous surface (1 m) temperature and salinity and relative fluorescence for return run along transect, March 11-12, 1977.

C. Chlorophyll *a* extracted concentrations at selected stations along the transect. March 11-12, 1977.

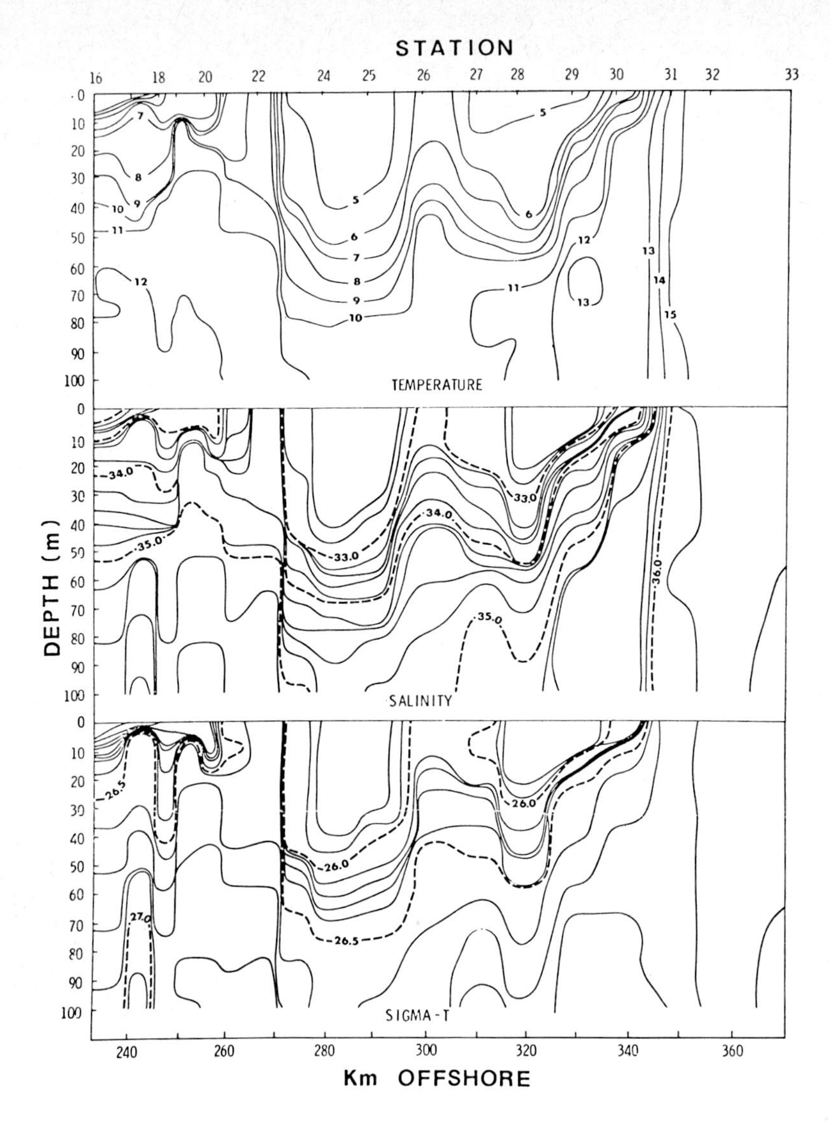

Figure 5. Hydrographic properties along transect, March 11-12, 1977.

profiles. Station 22, by contrast, is relatively well mixed to about 40 m with temperature and salinity high enough to correspond to that of slope water. A sharp prograde front is located between stations 22 and 23, seaward of which temperature and salinity again drop to the range of stations 16-21 but with lower vertical stability than at those stations. Between stations 29-32 another intense retrograde front was encountered, but with no observed synergistic effect on relative fluorescence. Temperature and salinity beyond station 31 were the highest encountered across the transect and suggested a strong admixture of Gulf Stream and slope water.

Figure 6 is a T/S diagram constructed from data from four selected stations across the transect (2, 16, 22, 26). Station 2 is outside Halifax harbour and is clearly representative of coastal water. Station 16 is in the inner frontal region and T/S properties show rapid vertical transition from surface coastal water into warmer, deeper, more saline slope water. Station 22 values corresponds to the 10°C water of Figure 4 and clearly indicate slope water, while station 26 data more closely resemble those of station 16 than at station 22.

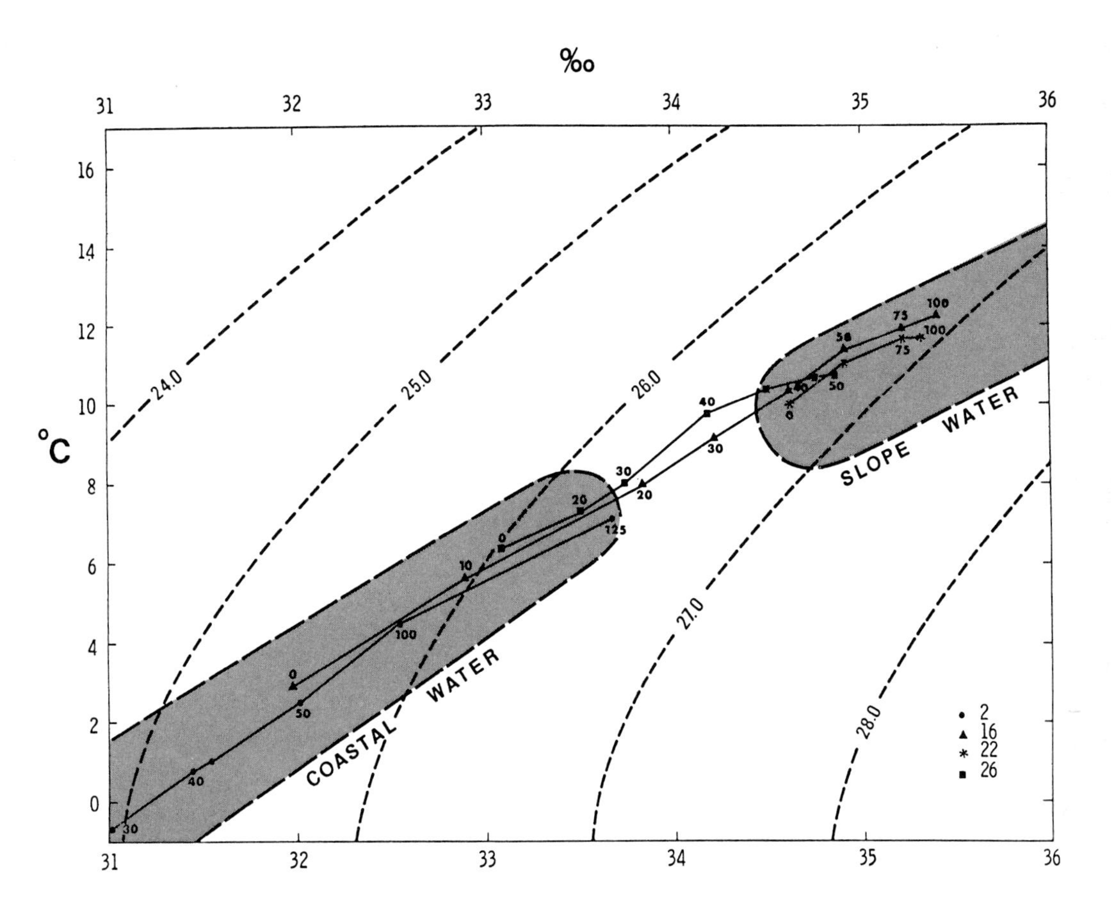

Figure 6. T/S diagram constructed from data of stations 2, 16, 22, 26, March 11-12, 1977.

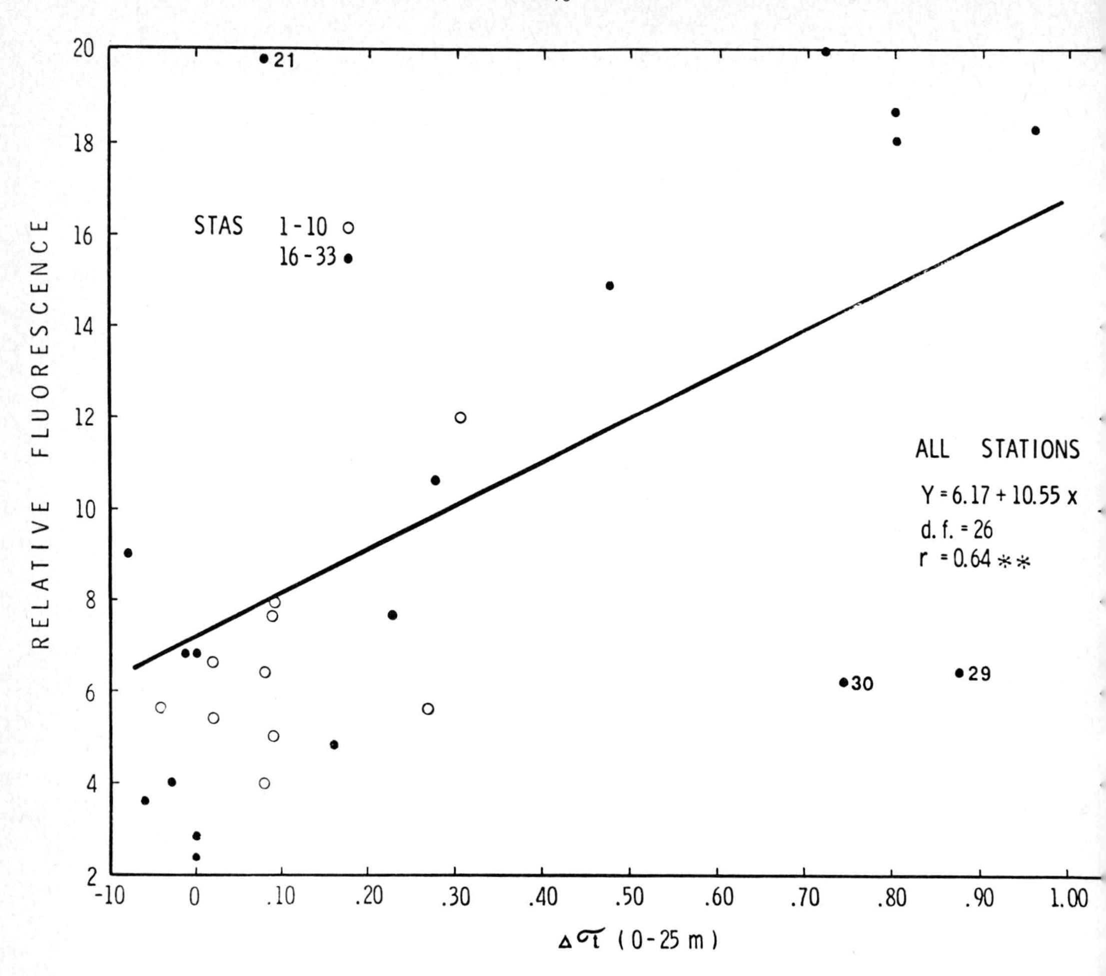

Figure 7. Regression plot of relative fluorescence versus 0-25 m density contrast (ΔσT), March 11-12, 1977.

On this cruise, station 22 water was formed of an intrusion or meander of slope water bounded with coastal water on both sides (16-21 and 23-30). Towards the end of the transect, water has been modified by mixing with water suggestive of a Gulf Stream source. Infrared NOAA-5 satellite photographs of this area at the time of the cruise support this interpretation.

This cruise in early March 1977 was made approximately one month prior to the spring flowering. At that season, intense vertical mixing was underway and nutrients were everywhere abundant. The onset of the spring flowering usually occurs when incident insolation has increased sufficiently at depth and when a reduction of the depth of the mixed layer occurs through surface stabilization from heat storage. During this period of observation, it appeared in the inshore frontal region that a reduction in the mixed layer depth was a consequence of mixing of the two water masses. The result is an apparent increase in light available to those phytoplankters which live in that shallow stratified area. Nutrients are abundant and light is sufficient as long as mixing is significantly reduced. Figure 7 supports this hypothesis by showing a correlation between observed relative fluorescence and $\Delta\sigma t$ between 0-25 m.

Conclusion

In general, in the winter months when nutrients are abundant and light is limiting, it is possible for phytoplankton growth to proceed at an accelerated rate in the localized vicinity of the Nova Scotia shelfbreak front. An important but as yet unknown fact concerns the duration of this phenomenon. Is this a typical winter condition or have we been fortunate to observe a short-lived phenomena on two consecutive years? The answer to this question remains to be answered. Equally important to investigate is the mechanism that permits relatively high phytoplankton growth to occur in the shelf/slope frontal region during the summer and late fall when nutrients are in short supply in the euphotic zone and thus limiting growth.

References

Fournier, R. O., J. Marra, R. Bohrer, and M. Van Det. 1977. Plankton dynamics and nutrient enrichment of the Scotian shelf. J. Fish. Res. Bd. Canada, 34(7):1004-1018.

Grant, D. A., and P. R. Rygh. 1973. Surveillance of fishing activities in the ICNAF areas of Canada's east coast. Canada Department of National Defence, Commander Maritime Command, MC/ORB report 3/73.

Uda, M. 1959. Seminar 2. Water mass boundaries - "siome" frontal theory in oceanography. Fish. Res. Bd. Canada. Manuscript report section 51 (unpublished).

9. HEADLAND FRONTS

ROBIN D. PINGREE, MALCOLM J. BOWMAN
AND WAYNE E. ESAIAS

Introduction

Headland fronts are formed in association with flow around headlands, promontories, banks, shoals and islands. Insight into headland frontal characteristics is best gained by considering frontal development in response to periodic flow past a headland rather than any steady advective flow which may not, in general, be a common feature of the non-tidal circulation. The main feature, therefore, that distinguishes this classification of tidal fronts from shallow sea fronts (see section 5) is their short characteristic length and time scales. Typically, headland fronts may be expected to go through a generation and dissipation phase within a tidal period. Consequently, if the dissipation is complete, the length scales for headland fronts will not exceed a tidal excursion. In other respects they exhibit characteristics common to other surface fronts (e.g. surface convergence, frontal jets, foam lines, changes in sea-state, color variations, etc.). Recent observations of fronts and flow patterns associated with headlands have been made by Bowman and Esaias (1977) and Pingree and Maddock (1977).

Effects of Headlands on Tidal Flow.

Fronts around promontories arise as a result of the influence of coastline configurations on the gross tidal flow. This invariably leads to:

(i) increased tidal streaming off headlands and weaker tidal velocities in adjacent bays,

(ii) large local variations in current phase resulting in jet-like flow off headlands in response to the advanced tidal stream phase of inshore waters,

(iii) a local lowering of mean sea level in the neighborhood of the headland due to curvature of the flow around the headland,

(iv) residual circulation patterns associated with headlands and promontories as vorticity generated at the headland is advected by the tidal flow and transferred to the mean circulation (see Figs. 1 and 2).

Frontal Development Associated with Headlands.

These various aspects of tidal flow lead to the development of fronts that are related to local geomorphological features, and which may result from a number of causes. In the first instance the strong tidal streaming in the neighborhood of headlands will cause a local minimum in the value of the h/u^3 stratification parameter (see section 5). Therefore, if stratified conditions exist offshore an abrupt transition to well mixed conditions may take place in the vicinity of the headland.

Even in the absence of offshore stratification, frontal characteristics may develop as a consequence of the residual tidal flow generated in the neighborhood of the headland. Fresh, warmer (colder in winter) water from adjacent bays will be drawn seaward off the headland where it will lie alongside water with properties more characteristic of offshore regions. Increased horizontal shear and jet like flow associated with headlands will increase the horizontal gradients between the different water masses and frontal characteristics will develop in response to the enhanced density gradients (see Figs. 1 and 2)

The nonlinear advection terms in the equations of motion (e.g., $u\partial_x v$ term in equation 2, p. 3-1) are mainly responsible for the local lowering of mean sea level in the neighborhood of headlands. This is a consequence of the well-known effects that when the streamlines converge, the surface elevation falls (Bernoulli's Theorem) and that when streamlines curve, the normal accelerations are balanced by a slope of the sea surface (centrifugal effect).

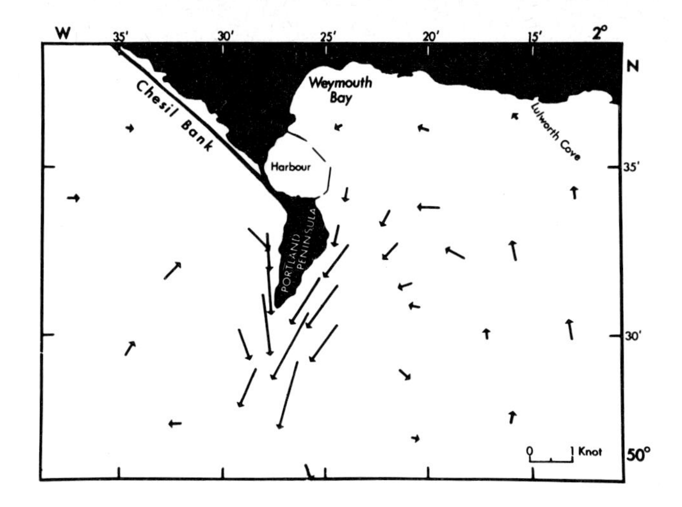

Fig. 1. Residual circulation (currents averaged over one semi-diurnal tidal cycle) of the water in the neighborhood of Portland Bill (south coast of Dorset, England) obtained from *current measurements* in the region. (These values are greater than those shown in Fig. 2 since the former refer to surface conditions at spring tides, whereas the values shown in Fig. 2 represent mean values for the whole water column during average tidal conditions. (Figs. 1 and 2 from Pingree and Maddock, 1977)

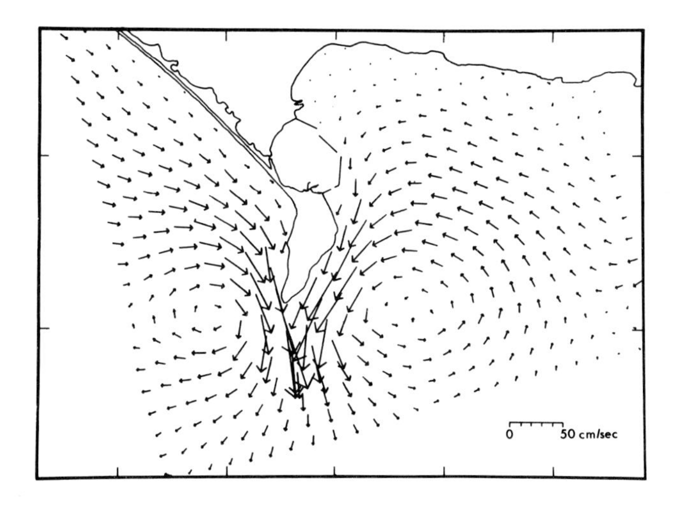

Fig. 2. Residual circulation of the water in the neighborhood of Portland Bill derived from a *numerical model* showing residual flow off headlands and the associated return eddies. For spatial clarity only one quarter of the derived current values have been plotted.

Both these effects will occur off headlands and since they act in the same way when the current flows in either direction a net depression of sea level occurs. In addition, Coriolis effects on the mean circulation, and other non-linear terms in the equations of motion contribute to local mean sea level changes.

The mean sea level off Portland Bill is depressed by 16 cm with respect to Weymouth Bay (Fig. 3) and a secondary circulation favoring upwelling may occur at the low pressure center.

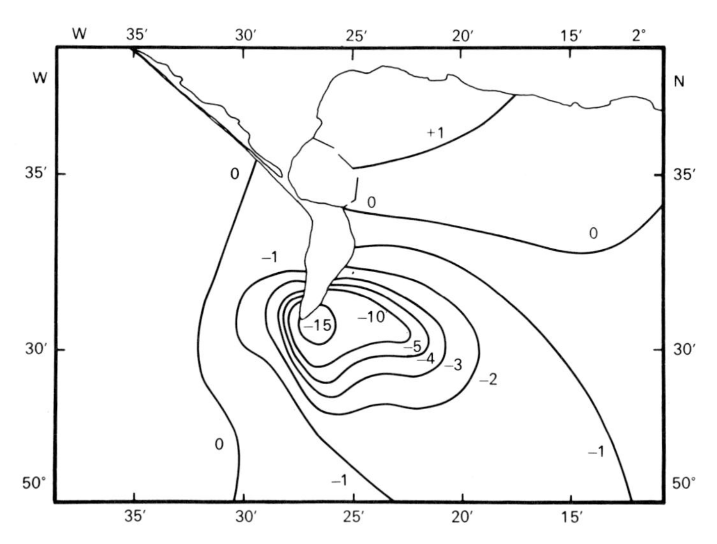

Fig. 3. Contours (in cm) showing the local lowering of mean sea level off Portland Bill. The reference level is taken as the mean sea level at the western entrance to the English Channel.

An example of a jet like flow around a local promontory due to increased streaming and local current phase variations is shown in Figs. 4-11. The jet has a remarkable subsurface core flowing at $\sim$70 cm sec^{-1} and clearly defined cross stream velocities. Surface contours of chlorophyll *a* in the region are shown in Fig. 12.

Future Work

A better understanding of headland fronts could be gained by further work defining the importance of vorticity transfer in frontogenesis, and to the more complete elucidation of the flow field. The relative rapid and repeatable generation and lifetime of this frontal type make the experimental work much more tractable than other, larger fronts.

Nutrient mixing and upwelling associated with headland fronts may be important to the local primary productivity (i.e., phytoplankton). In addition, the accumulation of buoyant detritus and neuston, and the aggregation of upward migrating zooplankton, may serve as a food source for nektonic species. Support for this conjecture is found in the frequent use of headland fronts as fishing sites by both sport and commercial fishermen.

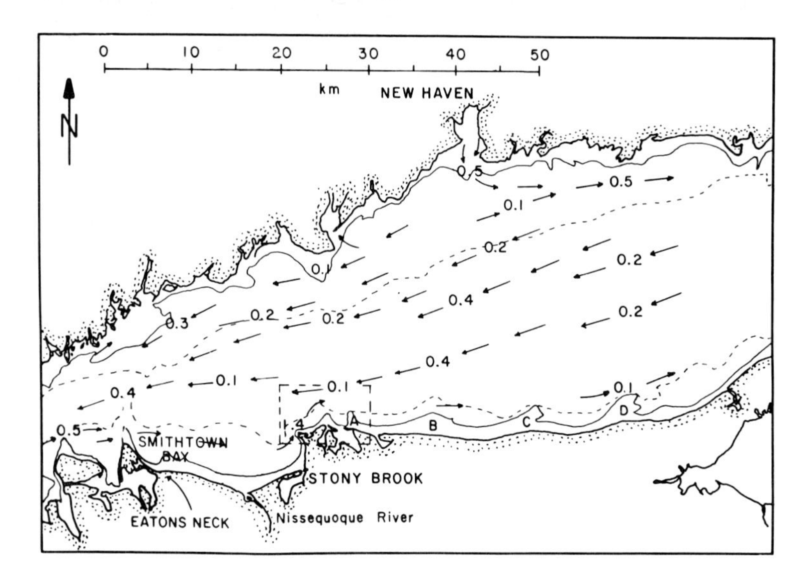

Fig. 4. Long Island Sound central basin. The inset represents the study area shown in Fig. 5. The solid bathymetric line is 20 feet; the dashed line is 60 feet. Arrows delineate tidal current direction; numerals, speed in knots. Note the 1.4 knot tidal jet around Crane Neck (see Fig. 5). Local time is "slack"; ebb begins at "The Race" (from Bowman and Esaias, 1977).

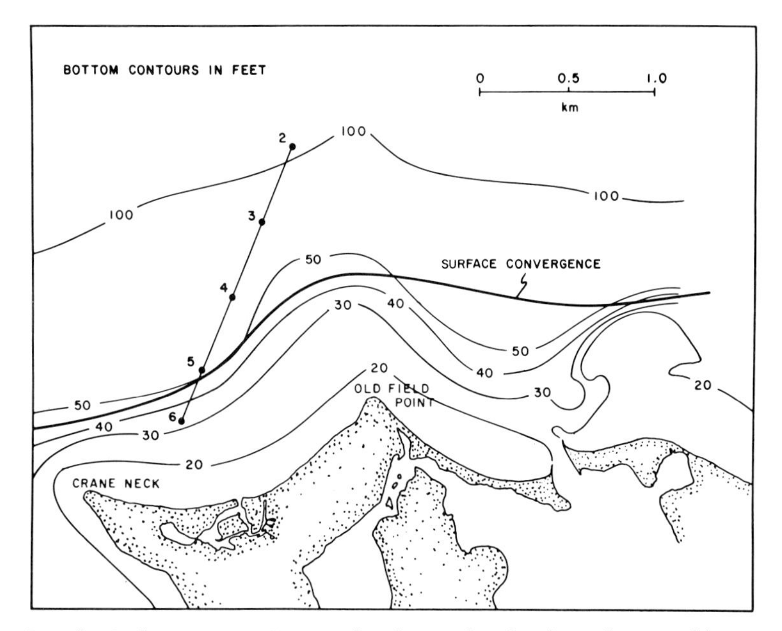

Fig. 5. Local study area in Long Island Sound, showing the sampling transect, and mean surface convergence position relative to the bottom contours. Depth contours are in feet (from Bowman and Esaias, 1977).

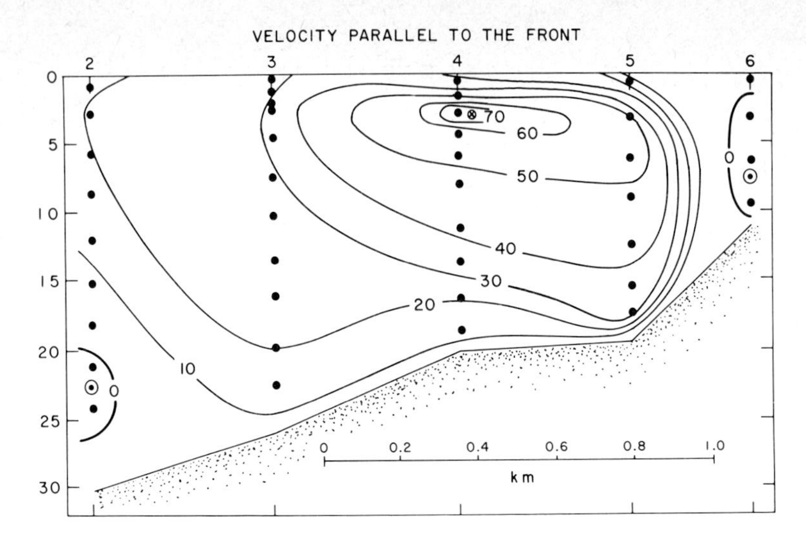

Fig. 6. Tidal current velocity (cm sec^{-1}) tangential to the surface front at station 5 (60° T) showing frontal jet at local time "slack: ebb begins at The Race" (Fig. 4). Depths are given in meters (from Bowman and Esaias, 1977).

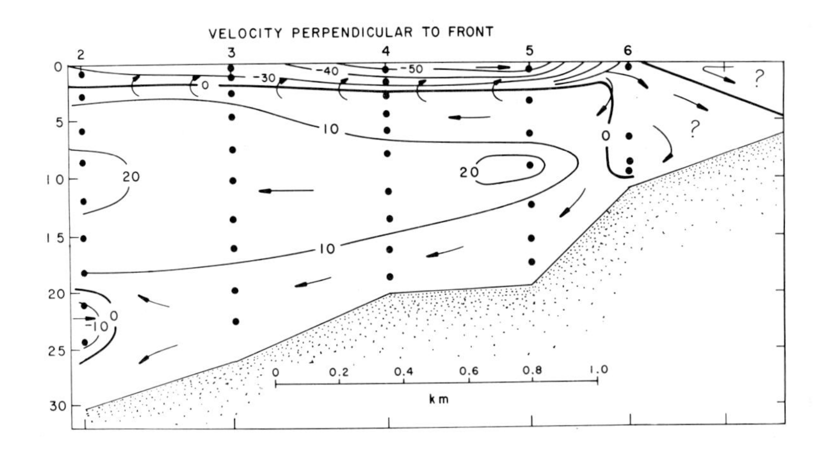

Fig. 7. Tidal current velocity (cm sec^{-1}) perpendicular to the surface front at station 5. The slope of the upper frontal interface inshore of station 6 is inferred from the density field. The arrows indicate streamlines (from Bowman and Esaias, 1977).

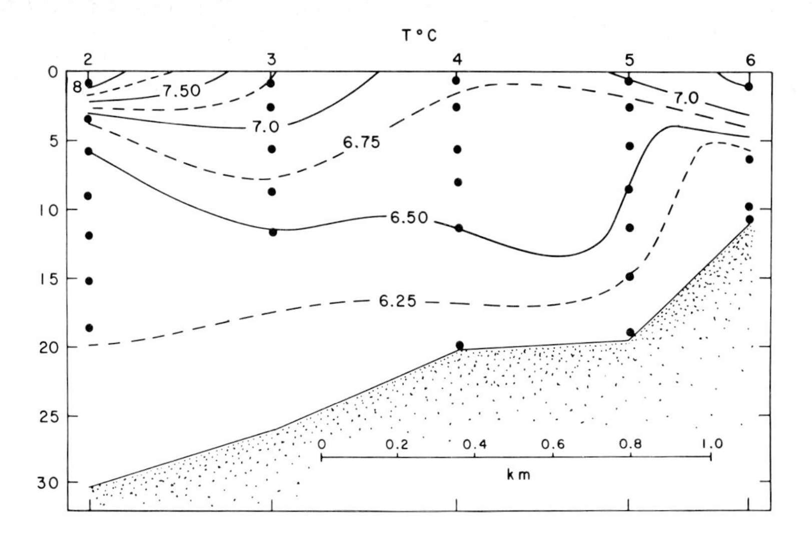

Fig. 8. Vertical temperature (C) section across transect shown in Fig. 5 (from Bowman and Esaias, 1977).

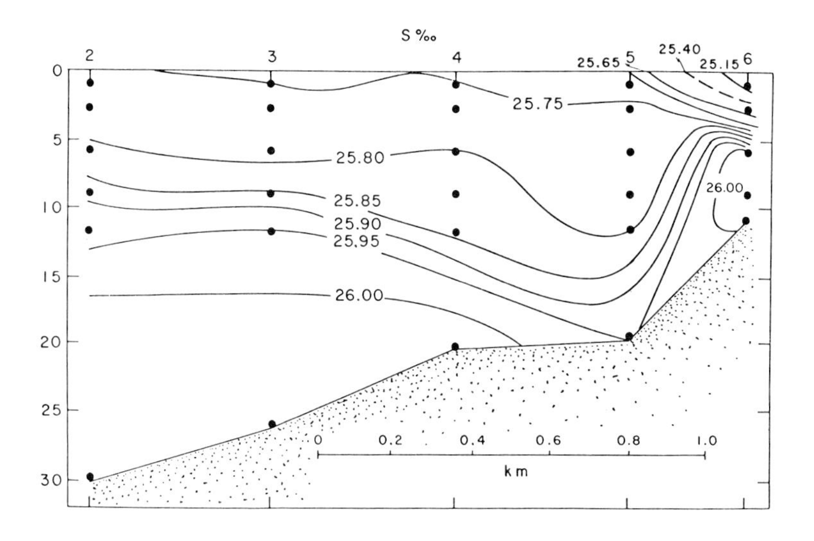

Fig. 9. Vertical salinity ($^{o}/oo$) section across the transect shown in Fig. 5. Note the break in contour interval between stations 5 and 6. There is a suggestion of upwelled bottom water of S = 26 $^{o}/oo$ at station 6. The headland front slopes up to the right between stations 5 and 6. The section is complicated by a second front, sloping down to the right above the headland front. The latter front is due to entrainment of low salinity Nissequoque River and Stony Brook Harbor water around Crane Neck (see Figs. 4 and 5) on ebb tide (from Bowman and Esaias, 1977). Note the suggestion of upwelling at station 6.

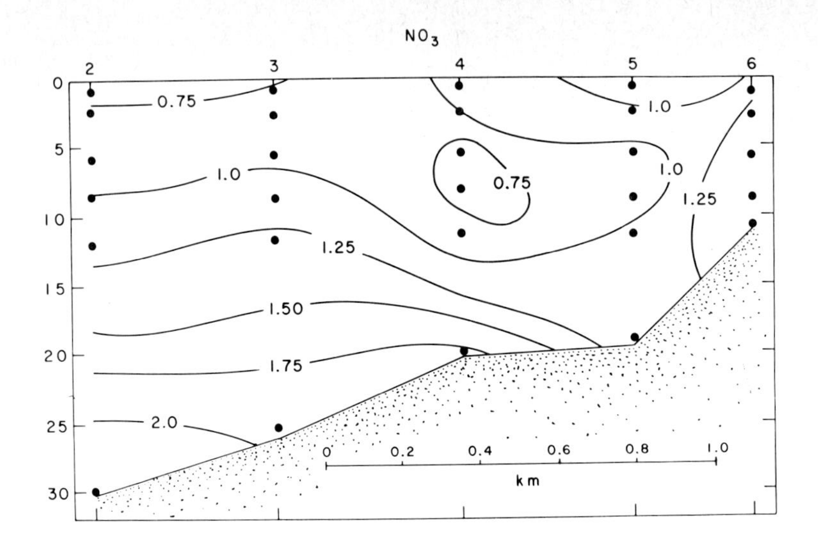

Fig. 10. Vertical inorganic nitrate (NO_3; μgm at l^{-1}) section across transect shown in Fig. 5 (from Bowman and Esaias, 1977).

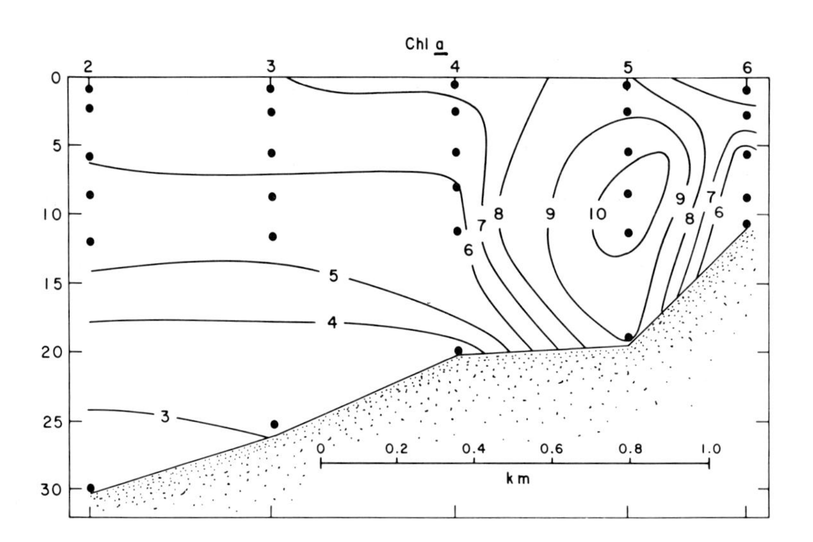

Fig. 11. Vertical chlorophyll *a* (mg m^{-3}) section across transect shown in Fig. 5 showing entrainment of high concentration phytoplankton into jet stream. The source of these phytoplankters is assumed to be Smithtown Bay (Fig. 12) (from Bowman and Esaias, 1977).

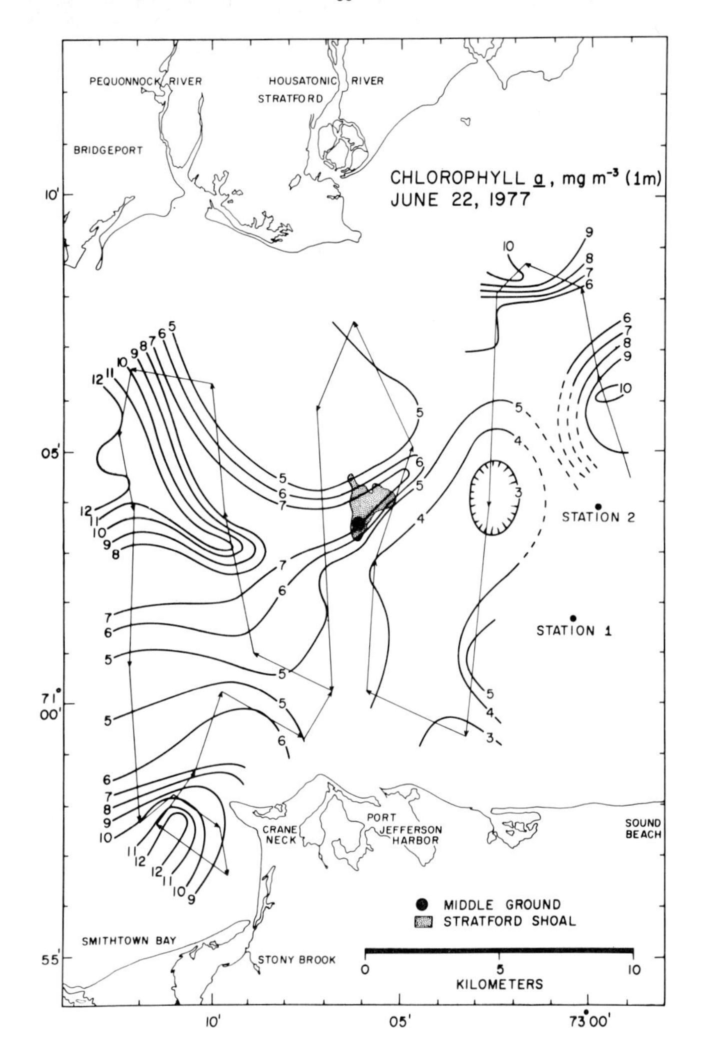

Fig. 12. Surface (1 m) chlorophyll *a* concentrations in Long Island Sound central basin June 22, 1977. Total travel time along cruise track was 9 hours. The Crane Neck frontal zone was sampled at the tidal phase shown in Fig. 4.

Whether such regions represent an *important* source for food requires further study.

An understanding of residual circulations and diffusion near the headlands is important in planning coastal waste disposal sites. Poorly placed outfalls may result in entrapment of materials in the eddies formed near the headlands (see Fig. 2). On the other hand, it may be possible to take advantage of the jet streams, (Figs. 2 and 6) to ensure a rapid seaward dispersal of wastes. Finally, a better understanding of the circulation patterns that develop near headlands may help to elucidate the physical principles that result in the formation of tidal banks or shoals, which have been observed to exist near the center of non-tidal gyres near headlands. These sediment traps are also areas of detrital accumulation with resultant effects on the benthos.

We wish to express our thanks to R. E. Wilson and H. B. O'Connors for the opportunity of gathering the chlorophyll *a* data shown in Fig. 12.

References

Bowman, M. J. and W. E. Esaias. 1977. Fronts, jets, and phytoplankton patchiness. In: "Bottom Turbulence: proceedings of the 8th Liege International Colloquium on ocean hydrodynamics." Edited by J. Nihoul, Elsevier Oceanography Series, 19, New York, 255-268.

Pingree, R. D. and L. Maddock. 1977. Tidal residuals in the English Channel." J. Mar. Biol. Assn. of U. K., 57, 339-354.

Pingree, R. D. and L. Maddock. 1977. Tidal eddies and coastal discharge. J. Mar. Biol. Assn. of U.K., 57, 869-875.

10. ESTUARINE AND PLUME FRONTS

MALCOLM J. BOWMAN AND RICHARD L. IVERSON

Introduction

In this chapter we discuss the properties and dynamics of small scale fronts with length and time scales of the order of the tidal excursion and period. Such fronts are commonly found in estuaries and shallow seas.

Although the mechanisms leading to frontogenesis for estuarine and plume fronts are different, viz., spatial variations in bottom generated turbulent stirring versus buoyant spreading, it is possible, to a first approximation, to apply the same cross frontal circulation dynamics to both categories.

Estuarine Fronts

Estuarine fronts are generally located parallel to the axis of the estuary, and can extend longitudinal distances of tens of kilometers. The frontal interface slopes downwards toward the center of the estuary or channel. They form in shoaling areas where tidally generated bottom turbulent stirring is sufficiently strong to break down any vertical stratification (Fig. 1).

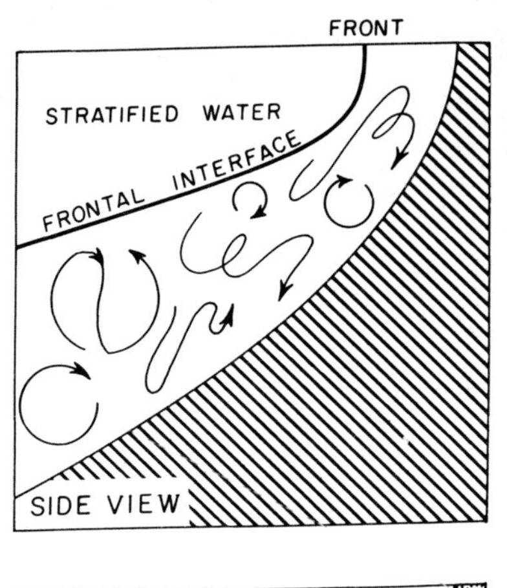

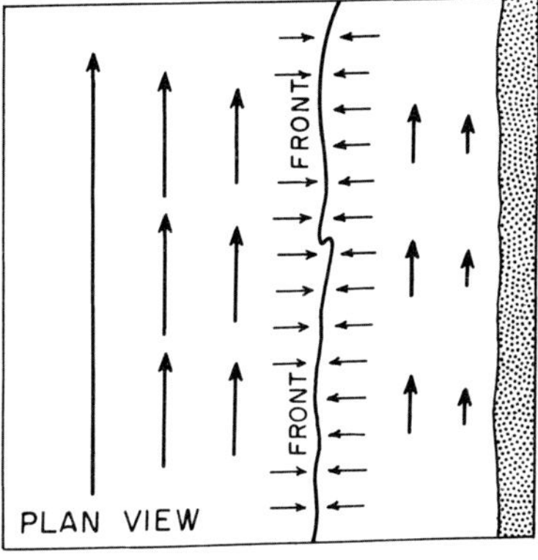

Fig. 1. Schematic diagram of an estuarine frontal zone. The curved streamlines represent turbulent bottom stirring.

The stratification which persists in deeper offshore water is due to surface buoyancy arising from lateral sources of runoff, an excess of precipitation over evaporation, or both. During the heating season insolation provides further buoyancy input through the sea surface.

Since the flux of energy from tidal streaming to bottom turbulence to water column mixing to cross frontal motion, are the same ones as explained in section 5 on shallow sea fronts, the h/u^3 parameter is also useful in locating estuarine fronts.

Lateral shear is important in maintaining these fronts. They are often observed to be stronger on ebb tide when faster flowing currents in deeper, offshore water advect lower salinity, upstream water past higher salinity shoal waters. This shear increases the horizontal density contrast across the front as well as sharpening the frontal interface *per se* (Fig. 2). On the flood tide, the fronts often weaken or vanish. Once formed, they may be advected away from their source of origin, and continue to exist until the difference in potential energy across the front has been converted into kinetic energy and subsequently dissipated.

The changes in properties across

estuarine fronts can be quite dramatic. In Delaware Bay, for example, surface salinity changes of $\sim 4^{o}/oo$ occur over a horizontal distance of a meter (Klemas and Polis, 1977). Strong convergence velocities $\sim$10-20 cm sec^{-1} associated with these fronts are very efficient in accumulating floating organic matter and detritus (Fig. 2).

Many estuarine fronts can be spotted from satellite or aircraft by a distinct color contrast across the front, attributed to differences in suspended particulate matter concentrations such as sediments or phytoplankton (Fig. 3). The foam and color boundaries are often located parallel to each other but separated by a few meters (Fig. 4).

Szekielda et al. (1972) found heavy metal concentrations three orders of magnitude higher than background values in frontal zones along Delaware Bay for chromium, copper, lead, mercury, silver, and zinc. Klemas and Polis (1977) also graphically illustrated how convergent tidal currents in Delaware Bay capture and advect oil slicks over periods as short as one hour.

Plume Fronts

In the region where the discharge of a river mixes with the receiving saltwater body, plume fronts often form. Depending on the rate of discharge and the geomorphology of the region, this mixing may take place within an estuary, or directly within the coastal ocean. In the first case, for sufficiently high runoff, other distinct plumes can form where the estuary itself discharges into the coastal sea. In this case the plume will consist of lower salinity, rather than fresh water.

Whether the buoyant discharges are a consequence of elevated temperatures (e.g., effluents resulting from power plant cooling) or lowered salinity, (e.g., river discharge), the driving mechanisms are similar. The effluent spreads as a buoyant plume over the receiving water (Fig. 5). If it were not for entrainment from the underlying water and interfacial friction, the plume would expand indefinitely as a steady thinning sheet, provided, of course, that the source was continuous. However, interfacial friction leads to the retardation of buoyant spreading and the

Fig. 2. Estuarine front with foam line photographed from an aircraft at an approximate scale of about 1:20,000. This front was observed just north of the lightering area inside Delaware Bay, with the more turbid water, indicated by a lighter shade of gray, being on the Delaware side of the front (courtesy of V. Klemas and D. Polis).

Fig. 3. Estuarine front without foam line photographed from an aircraft at an approximate scale of 1:10,000 in the middle of the lower portion of Delaware Bay. Note the shear displacement of ship wakes as they cross the boundary (courtesy V. Klemas and D. Polis).

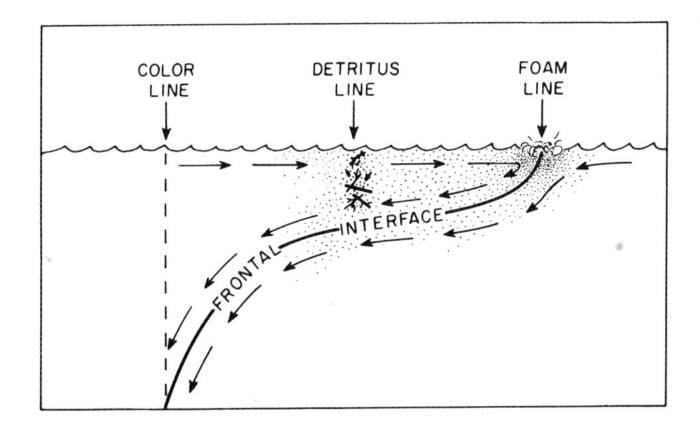

Fig. 4. Schematic cross section of a shallow front. Three boundaries are often visible:
i) The color front perceived to lie where the depth integrated upwelled light undergoes a distinct spectral shift in the region of rapidly descending isopycnals,
ii) The detritus line where large buoyant objects are trapped by oppositely directed currents at the surface and near the frontal interface,
iii) The foam line which is located at the surface convergence. Since the frontal slope may be $\sim 10^{-2}$, the three demarcations can be separated by several tens of meters. (After Klemas and Polis, 1977).

formation of sharp frontal boundaries at the leading edges of these plumes.

The driving mechanisms for frontogenesis are the horizontal pressure gradients established by the surface slope of the light water pool and the oppositely directed interfacial slope separating the plume from the underlying ambient water. The front will persist only as long as the two water masses are confluent, driven either by source discharge, or by a convergence of the ambient fluid which may be tidally or wind driven, or both. Unlike estuary fronts, the shallow draft of the plume fronts should free them from direct dependence on the bathymetry, as long as the water column is deep enough and bottom stirring is sufficiently weak.

If an estuary has a strong, reversing tidal flow, such as the Hudson under low discharge conditions, then coastal plume fronts will typically form only during estuarine ebb tide. As the flood tide begins in the estuary, reversing the surface slope across the plume, the fronts rapidly dissipate, an effect which can be observed many kilometers seaward from the mouth of the estuary (Bowman, in press).

The mixing dynamics of plume fronts differ from those of salt wedge circulation, since in plume frontal zones observational evidence suggests that surface water is entrained and mixed downward, while in the latter, entrainment is upwards and is attributed to breaking internal waves.

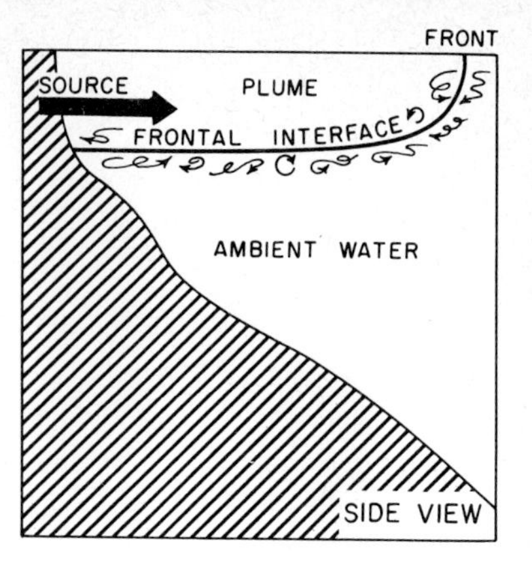

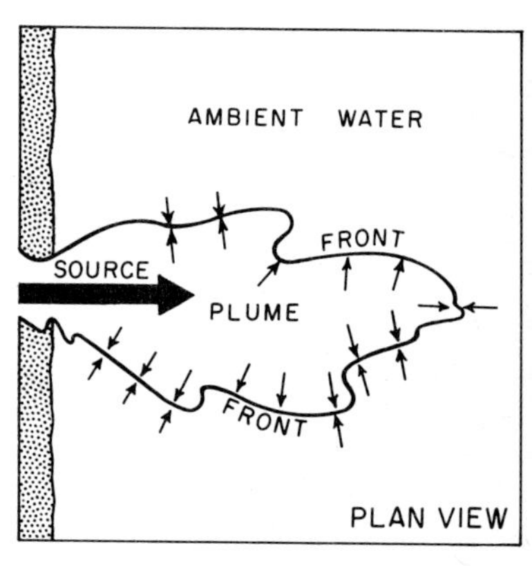

Fig. 5. Schematic diagrams of a plume front induced by buoyant spreading and interfacial friction.

Biological Implications of River Plumes

The spatial and temporal effects of river plumes upon coastal physical and biological processes depend on the magnitude and pattern of river discharge. Po River effects can be seen in the Adriatic Sea on a length scale of order 100 km (Revelante and Gilmartin, 1976). The Mississippi River (Riley, 1937) and Columbia River (Anderson, 1964) influence the nearshore environment on a length scale of order 400 km while the effects of the Amazon River can be seen on a scale of order 1000 km (Ryther, Menzel and Corwin, 1967).

In contrast to estuarine fronts, the time scale of river plume effects is of the order of weeks or months, as a function of seasonal discharge patterns and water column stability. River plumes can result in an enhancement of standing plankton biomass through nutrient enrichment of the photic zone followed by local plankton growth (Revelante and Gilmartin, 1976) or through advection of plankton over nearshore waters from a site of intensive growth near the river mouth (Malone, 1977).

The Amazon River contains lower concentrations of nitrate and phosphate than does the ocean water into which it flows (Ryther, Menzel and Corwin, 1967). There is enhanced diatom growth near the river mouth (Milliman and Boyle, 1975). A region of high biological productivity observed offshore between the coast of the Guianas and a lens of Amazon River water was attributed to effects of geostrophic divergence, partly related to the presence of river water (Ryther, Menzel and Corwin, 1967).

Nutrients discharged into the nearshore environment by the Columbia River are largely used in high production near the river mouth (Anderson, 1964). However, during the winter, freshwater runoff increases the area in which neritic diatoms flourish by stabilizing as well as enriching the nearshore water column (Hobson, 1966).

Temperature, salinity and color fronts associated with the Columbia River plume in summer have been described by Pearcy and Mueller (1970). High abundances and catches of albacore tuna are common in these frontal regions. It would appear that the albacore follow the oceanic plume front during their annual migration pattern (Pearcy, 1973).

Hudson River Plume

Some characteristics of the Hudson River plume are shown in Figs. 6-21. These diagrams were prepared from data gathered on three, one day cruises in the New York Bight Apex, during the low runoff summer season (August 13, 16, 19, 1976). Figures 6-10 are isometric perspectives of characteristics during cruise 1. The river discharge at the mouth during this period was $\sim$380 $m^3 sec^{-1}$ (the volume of fresh water in the plume $\sim 2.7 \times 10^8 m^3$, was equivalent to $\sim$8 day's river discharge). It can be seen that the plume spread as a very thin ($\sim$5m) lens over the coastal receiving waters, and exhibiting complex interdigitations. The prevailing wind during the preceeding 3 days was southwesterly; wind clearly strongly influenced the set of the plume, and it has drifted to the north and the east of its "usual" position along the New Jersey shore.

Strong fronts formed along the 29°/oo isohaline on Hudson estuary ebb tide particularly near the bases of the plume lobes, since the surface slopes of the light water pool were presumably greatest there.

Surface convergence currents inside the plume are very effective in sweeping flotsam into these fronts, and vast amounts of floating garbage can be traced at times for 30 km or more as filaments stretched along the front. Such debris includes oil, grease and tar balls, bottles, plastic devices and containers of all descriptions, paper, contraceptives, hospital refuse, abandoned boats, entrails, timber, fruit, toys, even cadavers.

The high chlorophyll *a* concentrations within the plume (Figs. 8, 13, 16, 19) reflected the high biomass supported by sewage derived nutrients (Figs. 9, 20) emanating from New York Harbor. Salinity fronts usually coincide with chlorophyll fronts, (Figs. 18, 19) but since the phytoplankton biomass is nonconservative, chlorophyll alone is not a good indicator

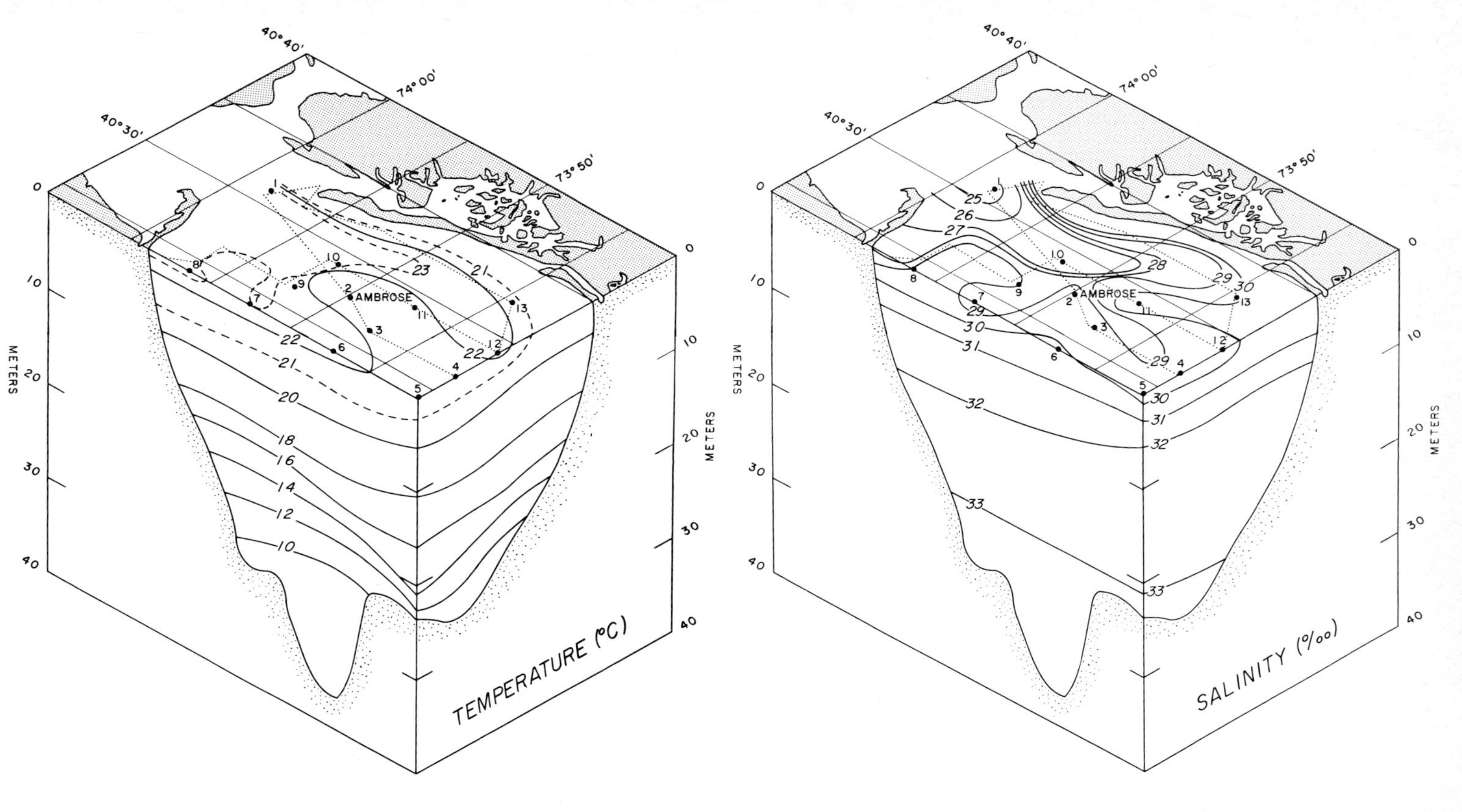

Fig. 6. Isometric perspective of temperature in New York Bight Apex, August 13, 1976.

Fig. 7. Isometric perspective of salinity in New York Bight Apex, August 13, 1976. Plume fronts commonly formed along the 29°/oo isohaline.

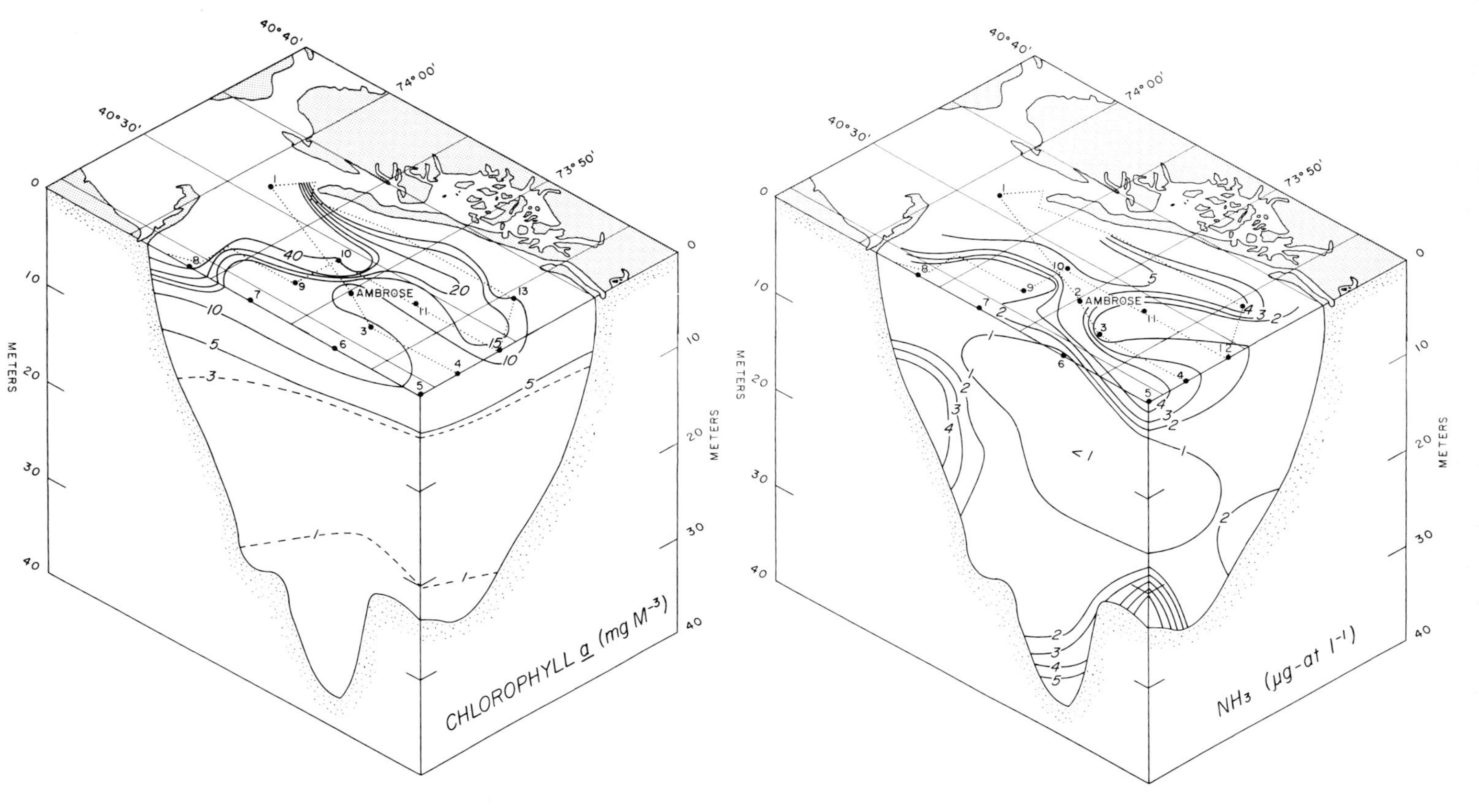

Fig. 8. Isometric perspective of chlorophyll *a* in New York Bight Apex, August 13, 1976.

Fig. 9. Isometric perspective of ammonia in New York Bight Apex, August 13, 1976. Ammonia is the principal sewage derived nutrient emanating from New York Harbor. High bottom concentrations at stations 8 and 5 reflected regeneration under the "normal" set of the plume along the New Jersey shore and in the dredge spoil dump site, respectively.

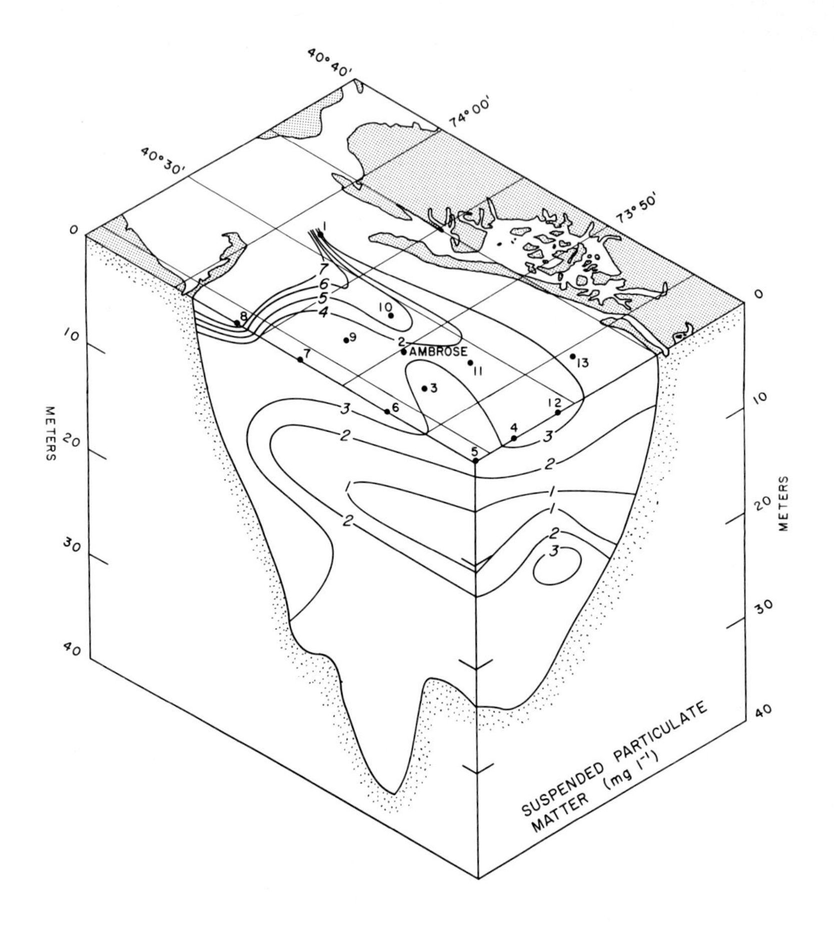

Fig. 10. Isometric perspective of suspended particulate matter concentrations in New York Bight Apex, August 13, 1976.

Fig. 11. Surface contours of temperature in Hudson plume, August 16, 1976. The solid zig-zag line is the cruise track.

Fig. 12. Surface contours of salinity in Hudson plume, August 16, 1976

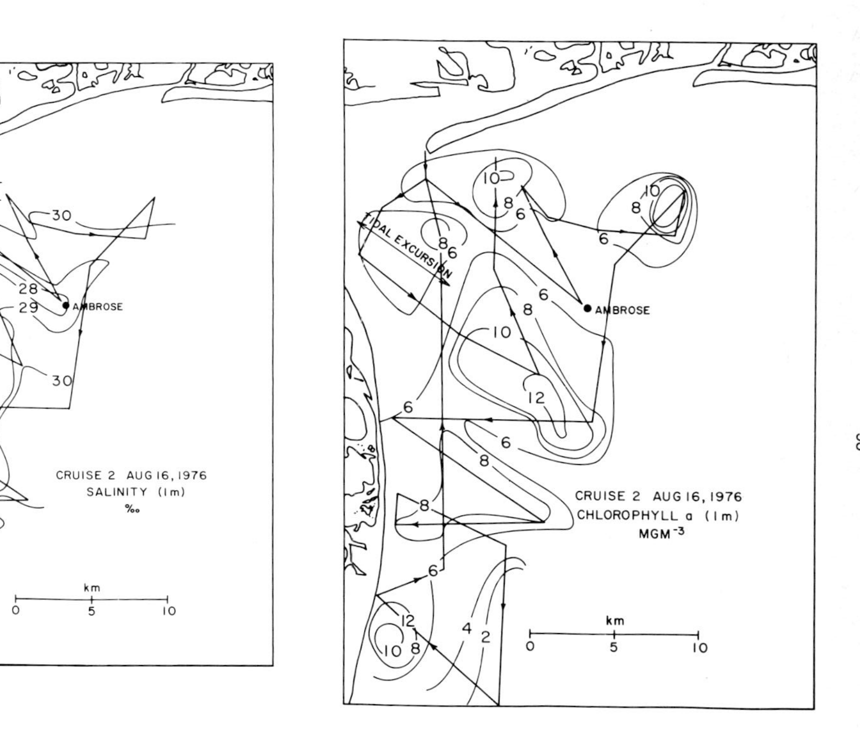

Fig. 13. Surface contours of chlorophyll a in Hudson plume, August 16, 1976

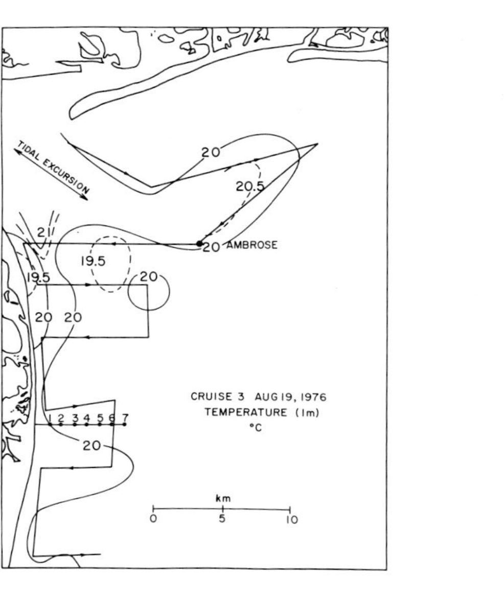

Fig. 14. Surface contours of temperature in Hudson plume, August 19, 1976

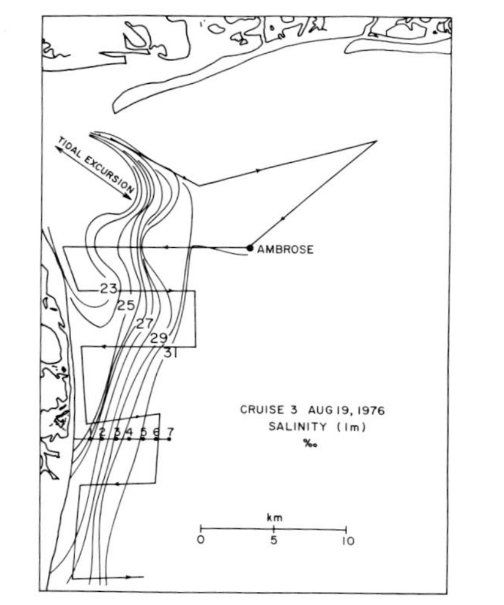

Fig. 15. Surface contours of salinity of Hudson plume, August 19, 1976

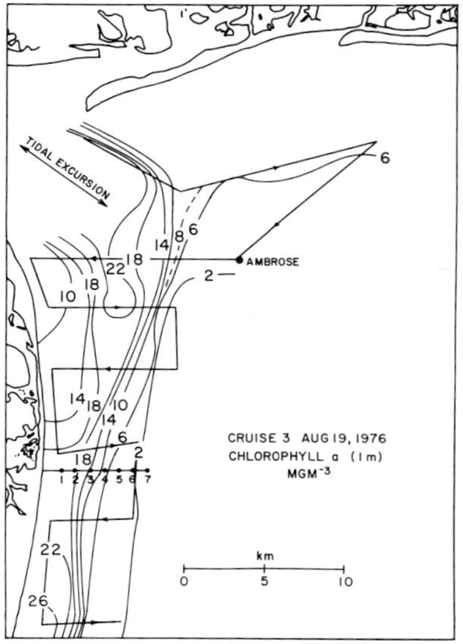

Fig. 16. Surface contours of chlorophyll *a* in Hudson plume, August 19, 1976

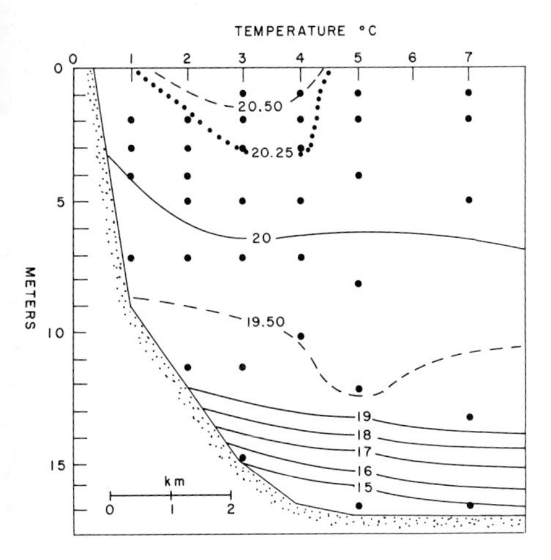

Fig. 17. Vertical temperature section through Hudson plume, August 19, 1976. See Fig. 14 for station positions.

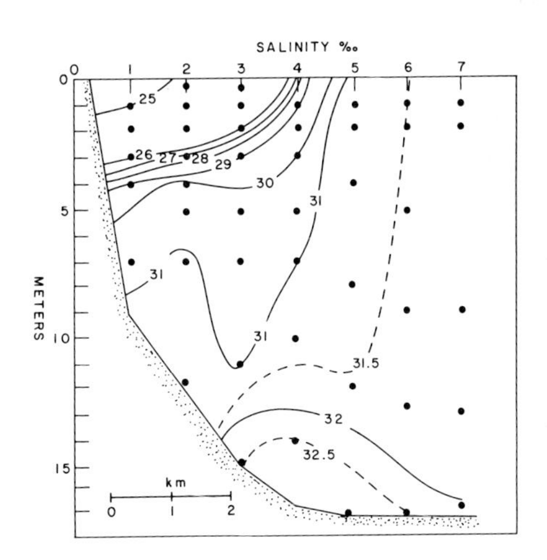

Fig. 18. Vertical salinity section through Hudson plume, August 19, 1976. See Fig. 15 for station positions.

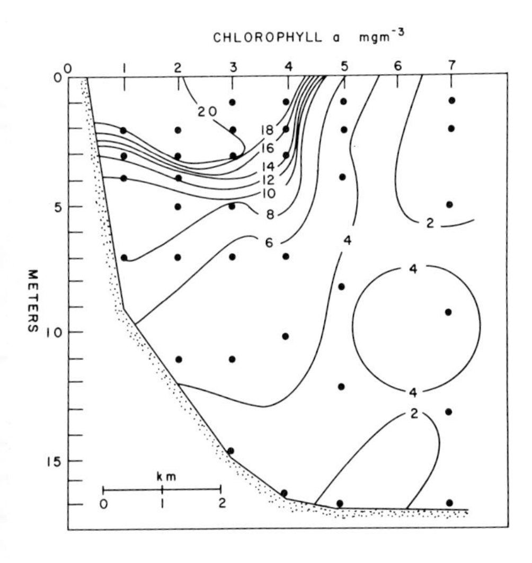

Fig. 19. Vertical chlorophyll *a* section through Hudson plume, August 19, 1976. See Fig. 16 for station positions.

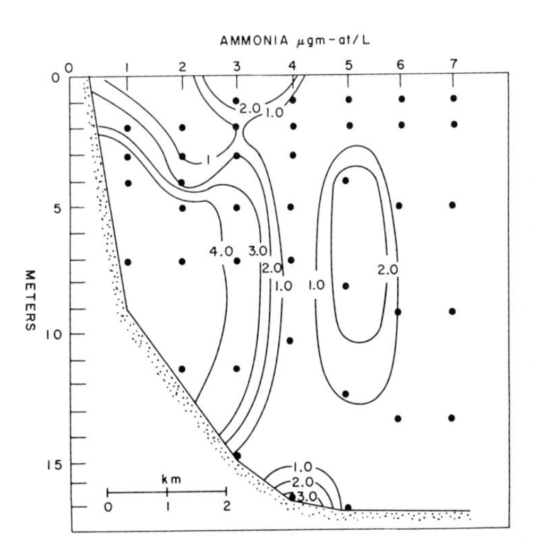

Fig. 20. Vertical ammonia section through Hudson plume, August 19, 1976. See Fig. 16 for station positions.

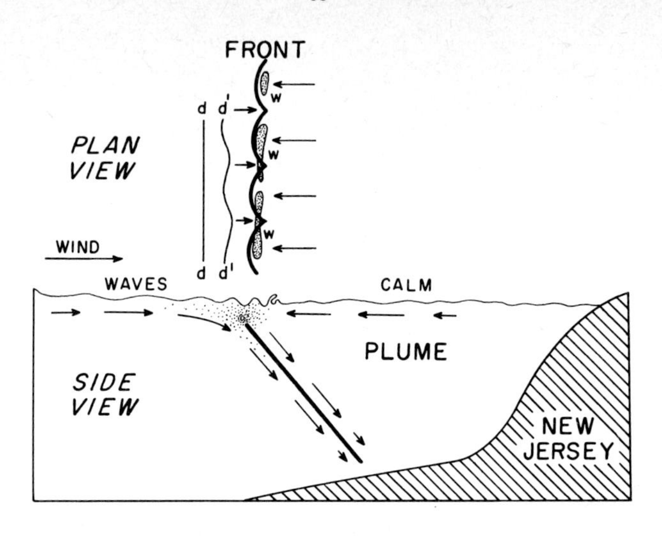

Fig. 21. A schematic picture of the front in cross section viewed to the south. The upper panel is a highly idealized plan view showing cusp like instabilities along the front, and collections of flotsam (w); d and d' represent undulations along wavefronts as surface gravity waves approach the frontal zone. (After Pingree, 1974)

or tracer of the plume (Figs. 12, 13, 15, 16).

Figures 11-16 illustrate the surface characteristics of the plume on August 16th and 19th (cruises 2 and 3). Characteristics measured during cruise 2 represented a transition state: the winds during this period were rotating from southerly through westerly to northwesterly, and the plume was drifting back to a southerly set. By cruise 3, the plume had reverted to a long ribbon along the New Jersey Coast. The winds during this period were northwesterly.

A vertical cross section of the plume is shown in Figs. 17-20. A sharp front at station 4 represented its outer limit. This front formed just before the section was taken and coincided with the beginning of ebb tide in the mouth of the estuary. The isohalines ($S=25-29^{o}/oo$) of Fig. 15 collapsed into a singular front which was subsequently followed after taking the transect, unbroken, northwards into the mouth of the estuary.

Discontinuities in properties were quite dramatic. Plume water was yellow-brown in color, with an oily smooth appearance. Oceanic water outside the front was coastal blue-green, with $\sim$1m surface gravity waves approaching the front, piling up, and breaking (Fig. 21). Again numerous filaments of flotsam delineated the front.

Frontal Dynamics

The following analysis closely follows that given in Officer (1976). The coordinate system is shown in Fig. 22.

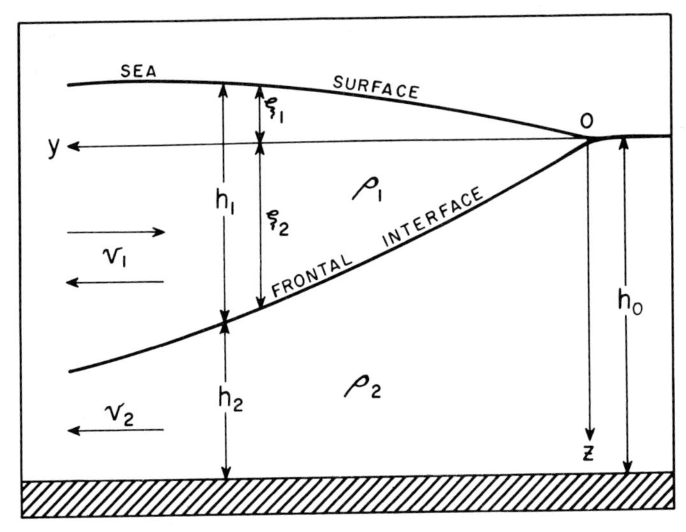

Fig. 22. Cross-frontal coordinate system with the surface front positioned at (0,0).

The y axis is in the cross-frontal direction, and z axis is positive downwards. The density in the surface layer is ρ_1, and in the lower layer is ρ_2 (both constant).

The surface elevation of the light water pool is ξ_1 (<0) and the depth of the frontal interface is ξ_2. The total depth of the upper layer is thus $\xi_2-\xi_1$. The total water depth is constant $=h_0$, the cross-frontal directed velocities are v_1 and v_2 in the upper and lower layers, respectively, and are measured with respect to the velocity of the frontal interface into the surrounding fluid.

Rotation is absent (i.e., Coriolis effects are neglected). This implies that the Rossby number $v_2/f\,\bar{y}$, where f is the Coriolis parameter and $\bar{y}$ the cross-frontal length scale, is much greater than unity.

Cross-frontal directed velocities and velocity gradients are assumed sufficiently small so that the nonlinear inertial terms may be ignored with respect to the internal friction terms. Any along-frontal directed velocity (such as tidal streaming) is considered constant and hence the lateral (cross-frontal) geostrophic force can be considered as a corrective term to the surface slope term. Internal friction is parametrized through a constant vertical eddy viscosity N_z.

The frontal interface is propagating to the right into the ambient fluid at a speed given from internal wave dynamics of $\left[g(\rho_1-\rho_2)\,\bar{h}/\rho_2\right]^{\frac{1}{2}}$ where $\bar{h}$ is the depth scale of the upper layer (Garvine, 1974b). The motion is steady relative to a coordinate system moving with the surface front at y=0; entrainment between the two layers is zero.

Thus, in steady state the equation of motions in the cross frontal direction reduce to:

$$\frac{\partial p_1}{\partial y} = \rho_1 N_z \frac{\partial^2 v_1}{\partial z^2}$$

$$\frac{\partial p_2}{\partial y} = \rho_2 N_z \frac{\partial^2 v_2}{\partial z^2}$$

and the hydrostatic equations:

$$p_1 = \rho_1 g(z-\xi_1)$$

$$p_2 = \rho_1 g(\xi_2-\xi_1) + \rho_2 g(z-\xi_2)$$

The continuity equations for the upper and lower layers are:

$$\int_0^{h_1} v_1 dz = 0$$

$$\int_{h_1}^{h_0} v_2 dz = q = \text{constant}$$

MODEL 1: no slip at sea bottom. The four boundary conditions are:

$$\frac{\partial v_1}{\partial z} = 0 \;,\; z = 0 \quad \text{(no surface stress)}$$

$$v_2 = 0 \;,\; z = h_0 \quad \text{(velocity is zero at the sea bottom)}$$

$$v_1 = v_2 \;,\; z = h_1 \quad \text{(no slip at the frontal interface)}$$

$$\frac{\partial v_1}{\partial z} = \frac{\partial v_2}{\partial z} \;,\; z = h_1 \quad \text{(stress continuous frontal interface)}$$

The solutions to these equations are:

$$v_1 = \frac{A}{3}(h_1^2 - 3z^2)$$

and

$$v_2 = \frac{B(h_0-z)(h_0-h_1)(z-h_1) - 2/3\, Ah_1^2(h_0-z)}{h_0-h_1}$$

where A and B are given by the solutions to the simultaneous equations:

$$B(h_0-h_1)^2 = -2/3\, Ah_1(3h_0-2h_1)$$

$$q = 1/6\, B(h_0-h_1)^3 - 1/3\, Ah_1^2(h_0-h_1)$$

Thus the model can be compared to observational data by measuring the transverse velocity relative to the front (e.g., using drogues) and the density field. The vertical eddy viscosity coefficient N_z is then given by

$$N_z = \frac{-3g\beta}{2A}\left[\frac{(h_0-h_1)^2}{3h_0^2-h_1^2}\frac{d\xi_2}{dy}\right], \quad \beta = (\rho_2-\rho_1)/\rho_2$$

Conversely, this equation can be integrated to yield the implicit frontal interface equation $\xi_2(y)$:

$$y = \frac{g\beta h_0^4 n^2}{12_g N_z}\left[1 - \frac{16}{9}n + \frac{14}{12}n^2 - \ldots\right]$$

where $n = \xi_2/h_0$ is the normalized depth. The shape of the front is parabolic near the surface, with a singularity at the origin (frontal slope becomes infinite), but gradually becomes more linear with depth (see Fig. 23).

Sketches of the velocity profiles are shown in Fig. 24. In essence the horizontal velocities are parabolic in form; the velocity at the interface is twice that at the surface, and in an opposite sense. The velocity is zero at $1/\sqrt{3}$ of the depth between the surface and the frontal interface (i.e., all motion is vertical). Maximum vertical shears are located at the interface and at the sea bottom.

Surface convergence is a feature of the model, and the frontal interface is a dividing streamline since no entrainment is allowed. At the front, water sinks and is drawn down the interface at a velocity, w, given by $w/v_1 = \partial\xi_2/\partial y$ (typically $\sim 10^{-2}$ cm sec^{-1}), several orders of magnitude greater than average open ocean values ($\sim 10^{-5}$ cm sec^{-1}).

Internal friction due to velocity shear sweeps the sinking surface water back away from the front against the horizontal pressure gradient which is always directed towards the front in the light water pool.

MODEL 2: Shallow fronts. In many estuarine and most plume fronts, the depth scale of the surface layer, ξ_2, is usually only a small fraction of the depth. In

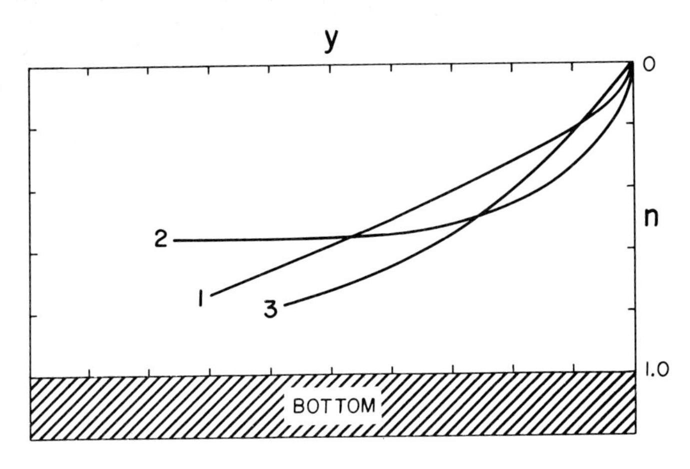

Fig. 23. Schematic diagrams of the frontal profile, for models 1, 2 and 3.

these cases we can use an alternate set of boundary conditions to replace the second two of model 1 and assume instead that:

$$v_2 = \text{constant}, \quad z > \xi_2$$

$$\rho_1 N_z \frac{\partial v_1}{\partial z} = k\ \rho_2 v_2^{\ 2}\ , \quad z = h_1$$

viz., the velocity in the lower layer is constant except for a thin boundary layer at the sea bottom, and interfacial stress is balanced by the bottom stress of the lowest layer as it flows over a rough bottom. The dimensionless drag coefficient k in the quadratic boundary layer stress term typically has a value of 2.6×10^{-3}.

The solutions of this model are:

$$v_1 = \frac{A}{3}\ (h_1^{\ 2} - 3z^2) \quad \text{as before, and}$$

$$v_2 = -A\ h_1^{\ 2}\ , \text{ but where } A = \frac{-9N_z\ \rho_1}{2kh_1^{\ 2}\ \rho_2}$$

$$\text{and} \quad B = 0\ .$$

These conditions lead to the frontal slope equation $\xi_2(y)$:

$$y = \frac{kg\beta\ \xi_2^{\ 4}}{36\ N_z^{\ 2}}$$

The frontal surface follows a quartic law, again with a singularity at the surface (Fig. 23), but with a more rapid leveling off with depth. Sketches of v are also shown in Fig. 24.

MODEL 3: Garvine's. Garvine (1974b) developed model 2 above a step further by arguing that the advective terms in the equations of motion are important, included interfacial entrainment, and applied a vertically integrated analysis. In addition he generalized the surface density to have a linear depth dependence and some specified monotonic horizontal increase determined by a dimensionless function $r(y)$ away from the front towards the source of the surface water. The bottom density remains constant.

The equations of motion and continuity then become in our notation:

$$v\frac{\partial v}{\partial y} + w\frac{\partial v}{\partial z} = -\frac{1}{\rho_2}\frac{\partial p}{\partial y} + N_z\frac{\partial^2 v}{\partial z^2}$$

$$\frac{\partial v}{\partial y} + \frac{\partial w}{\partial z} = 0\ .$$

Mass entrainment is parameterized by the empirical relation:

$$q_e = \pm\ Ev_2$$

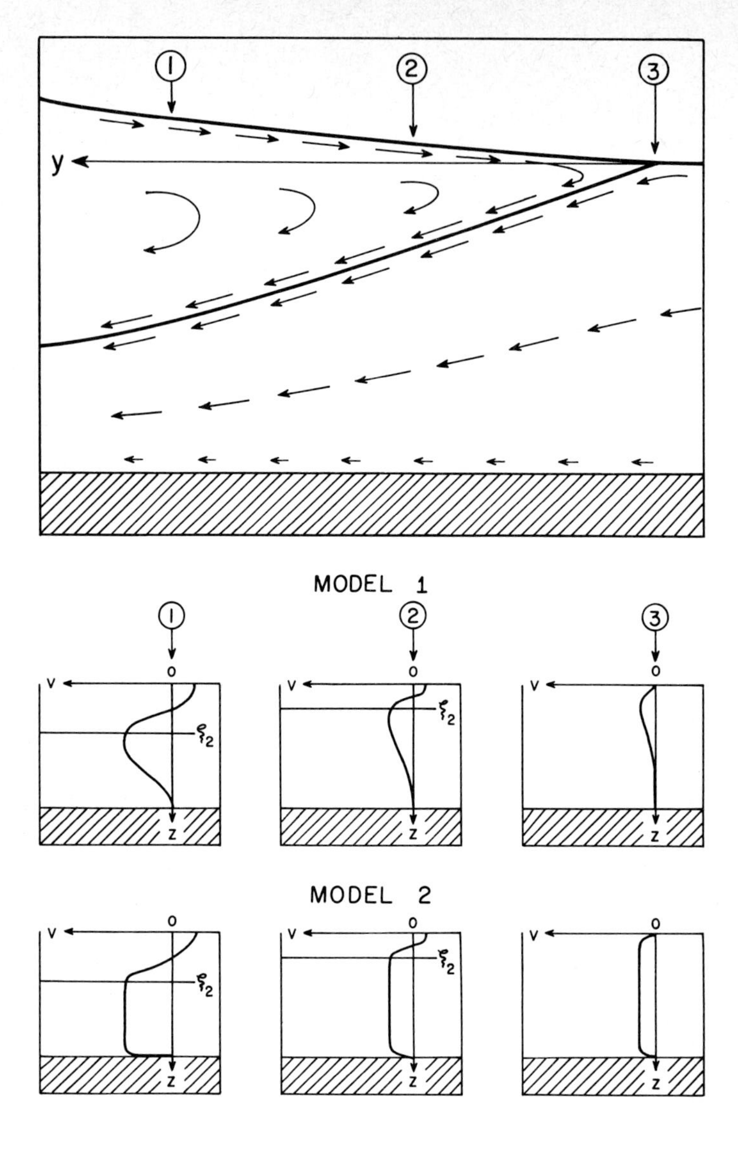

Fig. 24. Schematic diagrams of velocity streamlines and horizontal cross-frontal velocity profiles at three selected positions for models 1 and 2.

where $E \geq 0$, $<<1$ is a dimensionless entrainment coefficient, and the positive sign is for upward, and the negative for downward entrainment. E thus establishes the angle between the streamlines and the interface nearby.

Garvine investigated a special case of his model where $E=0$ and k = constant. He assumed the same solution for the velocity derived in the second model above, and showed that the interfacial slope is:

$$\frac{d\xi_2}{dy} = \frac{kv_2^2 - g\gamma\xi_2^2 r1/3}{v_2^2/5 + 2\, g\gamma r\xi_2/3}$$

where $\gamma = \left[\rho_2 - \rho_1(\infty)\right]/2\rho_2$

For ρ_1 constant, this equation reduces to

$$\frac{d\xi_2}{dy} = \frac{kv_2^2}{v_2 2/5 + g\beta\xi_2/3}$$

i.e., the frontal surface follows a parabolic law.

At the surface front, $d\xi_2/dy = 5k$. Hence one significant result of the model is that it removes the singularity of the frontal slope at the surface, and gives a simple relation between surface frontal slope and bottom stress.

It is of interest to compare the frontal slopes given by each model:

	Model 1	Model 2	Model 3
at $h_1 = 0$,	∞	∞	1.3×10^{-2}
$h_1 = 100$ cm	3.0×10^{-3}	2.5×10^{-2}	7.9×10^{-3}

(using $N_z = 8.5$ cm^2 sec^{-1}, $k = 2.6\times10^{-3}$, $\beta = 10^{-2}$, $v_2 = 50$ cm sec^{-1}, $h_0 = 10$ m, $g = 980$ cm sec^{-2}).

Finally by matching his theory to results obtained from Connecticut River plume fronts during a period of high runoff, Garvine concluded that the model gave reasonable agreement with experimental data, and that downward entrainment with $E \sim 4\times10^{-3}$ gave better agreement than upward entrainment.

Discussion. A great number of simplifying assumptions have been made in order to derive the analytical solutions discussed above. Although these results are useful for illustrative purposes, further studies are needed to investigate the validity of these assumptions.

In particular, fronts are seldom in steady state, the longitudinal flow may be confluent or diffluent, Coriolis effects may not be negligible, vertical eddy viscosities are certainly not constant in highly stratified flows, and the phenomena of interfacial friction and entrainment need to be investigated further.

More carefully planned experiments in the field are needed to improve our understanding of small scale fronts in shallow seas. These could include direct visual observations, using natural and introduced tracers, of mean circulation, shear, turbulence and entrainment to gain qualitative insight.

The only person we know who has the courage to regularly dive into strong convergence zones is Robin Pingree. He has observed violent motions in tidal fronts near Jersey off the north coast of France (Pingree, 1974). Pieces of weed were observed rotating violently at depth.

The most dramatic example of downwelling was when a small swimming jellyfish sank 20 m in 20 seconds. Observations of fluorescein dye ejected into the bottom mixed layer showed random motion in all directions with some evidence of vertical shear. Turbulence levels were so high that the experiment lasted only a few seconds.

Further controlled dye experiments could prove useful. A vertical line source of dye suspended from a buoy trapped in the surface convergence might shed light on frontal mixing, interleaving and entrainment. Larger scale dye experiments using continuous shipboard fluorometry to rapidly map vertical sections of concentration across the frontal zone, and over the tidal cycle could easily be designed.

Williams' acoustic velocity microstructure sensor (Williams and Tocho, 1977) might be mounted on a modified version of his SKIMP free falling temperature/ salinity profiler (Williams, 1974) to investigate velocity shear and Reynold's stresses in small scale frontal zones. Practical difficulties include upstream interference by the framework, and the impossibility of maintaining a level platform in violent turbulence.

Another tracer possibility is utilizing variance in plankton species composition across fronts as an indicator of water mass mixing. This method involves the use of rapid *in situ* pumping systems, and laborious species identification.

To summarize, we need to design multiple scale field experiments using new and imaginative techniques before we can hope to enlarge our insight into the elusive and fascinating dynamical properties of these fronts.

The Hudson plume study discussed in this chapter was supported by the MESA program of NOAA.

References

Anderson, G. C. 1964 . The seasonal and geographic distribution of primary productivity off the Washington and Oregon coasts. Limnol. Oceanogr., 9, 284-302.

Garvine, R. W. and J. D. Monk, 1974a. Frontal structure of a river plume. J. Geophys. Res., 79, 2251-2259.

Garvine, R. W. 1974b. Dynamics of small-scale oceanic fronts. J. Phys. Oceanogr., 4, 557-569.

Hobson, L. A. 1966. Some influences of the Columbia River effluent on marine phytoplankton during January, 1971. Limnol. Oceanogr., 11, 223-234.

Klemas, V., and D. F. Polis. 1977. Remote sensing of estuarine fronts and their effects on pollutants. Photogrammetric engineering and remote sensing, 43, 599-612.

Malone, T. C. 1977. Environmental regulation of phytoplankton productivity in the lower Hudson estuary. Est. Coastal Mar. Sci., 5, 157-171.

Milliman, J. D., and E. Boyle. 1975. Biological uptake of dissolved silica in the Amazon River estuary. Science, 189, 995-997.

Officer, C.B.,1976. Physical oceanography of estuaries(and associated coastal waters), John Wiley, N.Y. pp465

Pearcy, W. G. 1973. Albacore oceanography off Oregon. Fish. Bull., 71,

Pearcy, W. G., and J. L. Mueller. 1970. Upwelling, Columbia River plume, and albacore tuna. Proceedings Sixth International Symposium on Remote Sensing of the Environment. Univ. Mich., Ann Arbor, p. 1101-1113.

Pingree, R. D. 1974. Turbulent convergent tidal fronts. J. Mar. Biol. Assn. U.K., 54, 469-479.

Revelante, N., and M. Gilmartin. 1976. The effects of Po River discharge on phytoplankton dynamics in the northern Adriatic Sea. Mar.Bio. , 34, 259-271.

Riley, G. A. 1937. The significance of the Mississippi River drainage for biological conditions in the northern Gulf of Mexico. J. Mar. Res., 1, 60-74.

Ryther, J. H., D. W. Menzel and N. Corwin. 1967. Influence of the Amazon River outflow on the ecology of the western tropical Atlantic I. Hydrography and nutrient chemistry. J. Mar. Res., 25, 69-83.

Williams,III, A. J., and J. S. Tocho. 1977 An acoustic sensor of velocity for benthic boundary layer studies. In Bottom Turbulence, J.C.J. Nihoul, ed. Elsevier oceanography ser., 19, New York, 83-97.

Williams, III, A. J. 1974. Free-sinking temperature and salinity profiler for ocean microstructure studies. Ocean '74. IEEE international conference on engineering in the ocean environment, Vol. 2, Inst. of Elect. & Electronic Eng., Inc., N.Y., 279-283

Wright, L. D., and J. M. Coleman. 1971. Effluent expansion and interfacial mixing in the presence of a salt wedge, Mississippi river delta. J. Geophys. Res., 76, 8649-8661.

11. CROSSFRONTAL MIXING AND CABBELING

EDWARD P. W. HORNE, MALCOLM J. BOWMAN
AND AKIRA OKUBO

Introduction

In this chapter we investigate several dynamic processes in frontal zones which may be important under suitable conditions in effecting cross-frontal mixing and maintenance of sharp frontal gradients even in the presence of active diffusion. These are:

1) Ekman transport within internal boundary layers across geostrophically balanced fronts.
2) Crossfrontal interleaving and mixing across fronts where the horizontal density gradient may approach zero,
3) Cabbeling, which has been suggested as a mechanism for surface convergence and the removal of mixed water from the frontal interface in density balanced fronts,
4) Turbulent entrainment and downwelling induced by interfacial shear. This process appears to be especially active in small scale fronts, and has already been discussed in chapter 10.

Internal Ekman Boundary Layers

(The following discussion is based on conversations with G. Csanady)

We consider an inclined density front supporting a geostrophically balanced shear. An example of such a front is the Middle Atlantic winter shelfbreak retrograde front (Ch. 6). Internal turbulent Ekman layers form on both sides of the front, similar to the turbulent planetary boundary layer of the atmosphere.

An interfacial drag coefficient is thus predicted proportional to v_g^2/fN_z where v_g is half the magnitude of the geostrophic velocity difference across the front, f is the Coriolis parameter and N_z the vertical eddy viscosity. The interfacial stress drives an Ekman transport of the less dense fluid upward along the inclined front, and a downward transport of heavier fluid underneath the frontal interface.

This leads to a surface convergence in the denser fluid, and a surface divergence in the lighter fluid (see Fig. 9, Ch. 6 for an illustration). This upwelling of the lighter fluid may be responsible for some biologically important nutrient transport (Ch. 8).

The thickness of these Ekman layers will be quite variable depending on the geostrophic shear and density contrast but a typical value might be $\sim$15 m, with a volume transport of 0.3 m^2 sec^{-1} per unit length of the front.

Csanady also hypothesizes that intermittent release of the lighter upwelled boundary layer fluid might be responsible for the "bubbles" of shelf water observed in slope waters (Ch. 6).

Interleaving

STD profiles through oceanic thermohaline fronts often show multiple temperature and salinity inversions which have the nature of horizontal intrusions or fingers sometimes extending several km between the two water masses. The scale of these intrusions will depend on the level of turbulent mixing at the frontal interface, but they exist even when the density contrast across the front is undetectable.

Scales of motion. Three scales of motion can be identified. The first is the large scale, geostrophic flow associated with the water mass variations due to air-sea interactions. The second is the medium scale ($\sim$1-20 km) interleaving along horizontal or isopycnal surfaces, and the third is the microscale ($\sim$1 cm to 1 m) mixing of the two water masses driven by instabilities in the medium scale motion.

Hence, mass transfer across these fronts is accomplished by the production of intrusions by horizontal and isopycnal advection by medium scale motions of the existing large scale field. These intrusions are then dissipated by vertical microscale mixing such as salt fingering.

Dynamics of laminae. Stommel and Federov (1967) conducted pioneering STD investigations of thermocline intrusions in the Timor Sea (north of Australia) and also east of Mindanao in the Phillippine Sea. They described the intrusions as very thin laminae, extending from ~ 2 to 20 km in lateral extent, of thickness ~ 2 to 40 m vertically, and having lifetimes ~ 1 day.

They hypothesized that for small Ekman numbers, turbulent boundary layers form at the top and bottom interfaces of the laminae, i.e., the interior of the sheet is supposed to consist of a geostrophic current flowing parallel to the front, with cross-frontal directed internal Ekman flows near the upper and lower interfaces of thickness $z_e \sim (N_z/f)^{\frac{1}{2}}$ where N_z = eddy viscosity, and f = Coriolis parameter.

This cross frontal transport spreads the lamina of thickness ∂h into the other fluid at a typical speed of a few centimeters per second. This process can be envisioned as an equivalent diffusive spreading of the intrusion, with the rate of change of thickness of the lamina given by $\partial h/\partial t = (g' z_e/f)\ \partial^2 h/\partial x^2$ where g' is reduced gravity $g\Delta\rho/\rho$, $\Delta\rho$ is the deviation from the density ρ of the surrounding fluid, and x lies along the cross frontal direction. Thus, $K_H = g' z_e/f$ becomes an effective diffusivity of spreading.

On the other hand, if the layer is very thin compared to the Ekman layer then viscous dynamics predominates and the spreading equation becomes

$$\partial h/\partial t = \frac{g'}{12\ N_z} \frac{\partial^2}{\partial x^2} (h^4)$$

The effective diffusivity tends to zero at both large and small Ekman numbers, since for small values the layers are too thin to possess significant transport, while for large values the fluid is too viscous to flow rapidly.

Stommel and Federov evaluated the effective lateral diffusivity K_H at an intermediate Ekman number when it has the same value for both processes: viz., $K_H = 0.55\ g'h/f$. This value, given by the intersection of two asymptotic curves, is an approximation to the maximum effective diffusivity, and there is no reason to suppose that the real Ekman number corresponds to this maximum. Also, Stommel and Federov used a no slip boundary condition which underestimated the lateral transport.

The problem needs further attention, taking into account these factors and the double diffusive change in density as the layer intrudes.

Medium to microscale transfers. In steady state this horizontal cross-frontal advection of heat and salt is balanced by small scale vertical diffusion. As pointed out by Joyce (1977) the enhancement of cross-frontal heat transfer by interleaving is similar to that achieved in a car radiator. The thinner the layers, the greater the stirring and total area of contact, and hence the greater the potential for heat and salt exchange.

Joyce also suggested some criteria for assigning scale boundaries between the medium and small scales.

1) The medium scale motions are mainly horizontal in contrast to the microscale where velocities are vertical or random,

2) The T/S correlation is high for medium scale motions, and the variability is due to horizontal stirring of an existing large scale inhomogeneity. Small scale transfers are mainly vertical and can possess small T/S coherence,

3) There may be significant differences in the nature of the vertical wavenumber spectra of temperature and salinity between medium and microscales.

Vertical diffusion. In Fig. 1 T/S curves for profile numbers 14,15 and 18 of Station 62 are shown. Layer A of Fig. 5, Ch. 7, appears to be spreading along the 27.0 σ_T surface from right to left and there is some suggestion that the layer is gradually decreasing in density, although the instrument did not completely resolve the interfaces of the layers. Since these layers possessed sharp interfaces, this suggests that double diffusion may be occurring.

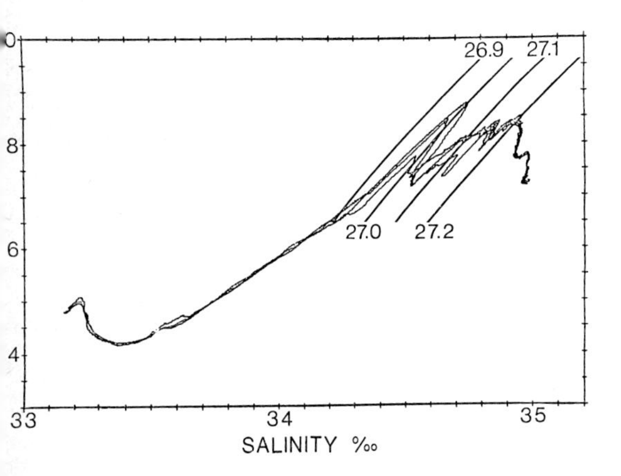

ig. 1. T-S curves for profile numbers 14, 15 and 18 of station 62 (see intrusion A, Fig. 5, Ch. 7).

If an interface consists of cold fresh water lying above warm salty water (a diffusive interface) then temperature is driving force for the instability, while, if hot salty water lies above cold fresh water (a salt finger interface) then salinity is the driving force for the instability.

Turner (1965) performed laboratory experiments to measure the heat and salt fluxes through a *diffusive interface* and Huppert (1971) fitted a curve to this data to find that:

$$F_T = 0.32\ (K_T)^{2/3}\ (\frac{\alpha g}{\nu})^{1/3} (\Delta T)^{4/3} (\frac{\beta \Delta S}{\alpha \Delta T})^{-2} \tag{1}$$

$$\frac{\alpha}{\beta}\ (1.85-0.85\ \beta\Delta S/\alpha\Delta T) F_T \qquad 1 \leq \frac{\beta\Delta S}{\alpha\Delta T} \leq 2 \tag{2}$$

$$F_S = 0.15\ \frac{\alpha}{\beta}\ F_T \qquad \frac{\beta\Delta S}{\alpha\Delta T} \geq 2$$

F_T is the heat flux in m C sec^{-1}, K_T is the thermal diffusivity, $\alpha = -\frac{1}{\rho}(\frac{\partial\rho}{\partial T})_{S,P}$ is the coefficient of thermal expansion, g is the acceleration due to gravity, ν is kinematic viscosity, ΔT the temperature change across the interface, ΔS the salinity change across the interface, $\beta = \frac{1}{\rho}(\frac{\partial\rho}{\partial S})_{T,P}$ and F_S is the salt flux in m^o/oo sec^{-1}.

Turner (1967) also measured the heat and salt flux through *a salt finger interface*. By fitting a straight line to the data in his paper:

$$F_S = 0.085\ (46 - 3.87\ \frac{\alpha\Delta T}{\beta\Delta S})\ (K_S)^{2/3}\ (\frac{g\beta}{\nu})^{1/3}\ (\Delta S)^{4/3} \tag{3}$$

$$F_T = r\ \beta/\alpha\ F_S \tag{4}$$

where K_S is the molecular diffusivity for salt, and r was estimated by Turner to be 0.56 for heat-salt fingers. Linden (1973) suggests that these measurements were made in the presence of turbulence and non-steady motions. He claimed that a value for r ~0.1, estimated from heat-sugar fingers, may be more appropriate.

The double diffusive fluxes just described can be parameterized by vertical eddy diffusivities for temperature, K_{VT}, and salinity, K_{VS}. To estimate K_{VT}, consider Fig. 2 illustrating a typical layer in the T and S profiles. Equation (1) can be used to calculate F_{T1} and (4) can be used to calculate F_{T2}. Then K_{VT} follows from

$$F_{T1} - F_{T2} = K_{VT} \frac{d\overline{T}}{dz} \qquad (5)$$

where $\frac{d\overline{T}}{dz}$ is the mean temperature gradient in the absence of the layers.

An estimate of K_{VS} can be obtained by using (2) and (3) to calculate F_{S1} and F_{S2} respectively. Then

$$F_{S1} - F_{S2} = K_{VS} \frac{d\overline{S}}{dz} \qquad (6)$$

where $\frac{d\overline{S}}{dz}$ is the mean salinity gradient.

These fluxes were evaluated for a number of interfaces shown in Fig. 5, Ch. 7 (details are given in Horne, in press). Average values of $K_{VT} = 8\ cm^2\ sec^{-1}$ and $K_{VS} = 3\ cm^2\ sec^{-1}$ obtained fall within the value of $28 \pm 38\ cm^2\ sec^{-1}$ obtained by Oakey and Elliott (1977) by applying the Osborn-Cox (1972) model to their microstructure measurements in slope water.

Lifetime and cross-frontal velocities The lifetime τ of the layer can be approximated as the time required for vertical double diffusive fluxes to annihilate the temperature and salinity anomalies, ΔT and

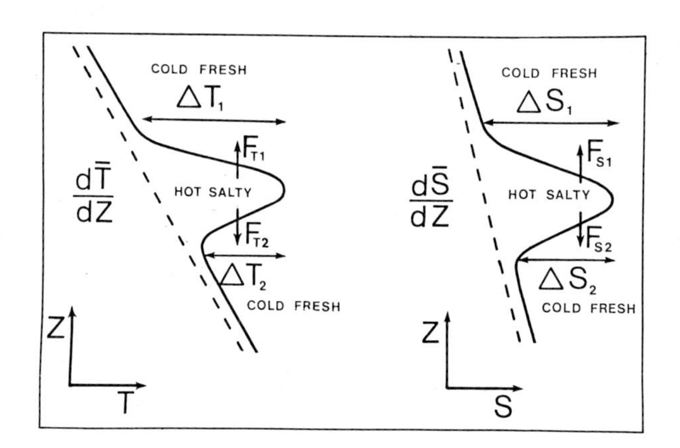

Fig. 2. A typical interleaving layer with heat and salt fluxes through its upper and lower interfaces.

ΔS, of the intrusion. For $\Delta T = 2.1$ C and $\Delta S = .5^{o}/oo$, (layer A, Fig. 5, Ch. 7) the lifetime $\tau \sim 30$ hours. The width of this intrusion is ~3 km which implies a cross-frontal velocity v of ~3 cm sec^{-1}.

Horizontal diffusion. An estimate of the horizontal heat flux F_{HT} is given by $F_{HT} = 0.5\ v\Delta T$, where ΔT is the temperature contrast across an interface. For $\Delta T = 0.6$ C, $F_{HT} \sim 9x10^{-3}$ m C sec^{-1}.

The effective horizontal diffusivity K_{HT} is estimated from $F_{HT} = K_{HT} \frac{\partial \overline{T}}{\partial x}$, where $\partial \overline{T}/\partial x$ is the mean horizontal temperature gradient.

Taking $\partial \overline{T}/\partial x$ at the front to be $8.3x10^{-4}$ C m^{-1} leads to $K_{HT} \sim 10^{5}$ cm^{2} sec^{-1}. A similar calculation for salinity using $\Delta S \sim 0.1^{o}/oo$ and $\partial \overline{S}/\partial x \sim 1.7x10^{-4}\ ^{o}/oo$ m^{-1} gives $K_{HS} \sim 0.9x10^{5}$ cm^{2} sec^{-1}. These values are consistent with the length scale of the intrusion $\ell \sim (K_{H}\tau)^{\frac{1}{2}} \sim 1$ km.

Voorhis et al. (1976) have investigated intrusions and mixing in the shelf-break front south of New England using neutrally buoyant floats and STD profiles. They found mean cross-shelf and vertical currents undetectable, but found horizontal displacements in the frontal zone consisting of clockwise orbital motions, roughly circular, having periods between the diurnal tide (24.84 hr) and the inertial period (18.7 hr). Vertical motions were more complex, dominated by both the above subinertial motions and at other times by semidiurnal and quarter-diurnal motions.

Frontal interleaving with vertical and horizontal scales ~10 m and 10-30 km respectively, were ubiquitous features during the June 1971 study. Voorhis et al. claimed vertical shear was effective in thinning the layers which had a lifetime ~3 days.

They estimated cross-frontal transports of heat and salt $\sim 2.5x10^{13}$ cal day^{-1} km^{-1} of shelf edge and $1.5x10^{10}$ gm salt day^{-1} km^{-1} respectively. These compare favorably to the estimates of $\sim 2.8x10^{13}$ cal day^{-1} km^{-1} and $\sim 4x10^{9}$ gm salt day^{-1} km^{-1} calculated for the Scotian shelf (Ch. 7).

Cabbeling

A common misunderstanding is that since frontal zones by definition are regions with sharp gradients of properties, they must therefore represent a barrier to mixing between the two juxtaposed water masses. Otherwise diffusion across the zone would soon dissipate the front and smooth out any horizontal discontinuities.

However, in reality, the opposite is true. The frontal zone is a region of *intensified* motion and mixing. Vertical velocities, for example, may be thousands of times larger than typical open ocean values. Turbulence levels can be intense (e.g., see Chapters 1 and 10).

How then is the front maintained in the presence of intensified mixing and diffusion of properties across the zone? Obviously some mechanism(s) is(are) needed to remove the mixed water away from the frontal interface to maintain its identity. Internal Ekman transports at frontal interfaces have already been discussed where light water "glides" up the interface and heavy water sinks under.

A second mechanism is cabbeling (from the German nautical terminology "Kabbelung" meaning a rippled or choppy appearance of the sea surface due to convergence).

Witte (1902) first pointed out that since the equation of state is nonlinear, a mixture of two water masses of dissimilar temperature and salinity but equal density, possesses a density greater than the components. Thus, he hypothesized that in a vertical frontal zone, a sinking would occur in the narrow mixed layer.

Neglecting internal friction, he calculated a typical buoyancy acceleration of ~.1 cm sec^{-2}, leading to a 100 cm sec^{-1} downwelling velocity over 400 m in depth. Although we now know that such velocities are greatly exaggerated (typically ·1 mm sec^{-1} or 10 m day^{-1}), Foster (1972) has developed a time dependent numerical model which suggests that from 8 to 50×10^6 m^3 sec^{-1} of Antarctic Bottom Water might be produced by the cabbeling instability process in the Weddell Sea. This flux plus the water that is modified over the Antarctic shelf area could then account for the yearly production of Bottom Water.

A linear cabbeling model. With reference to Fig. 3, we develop an

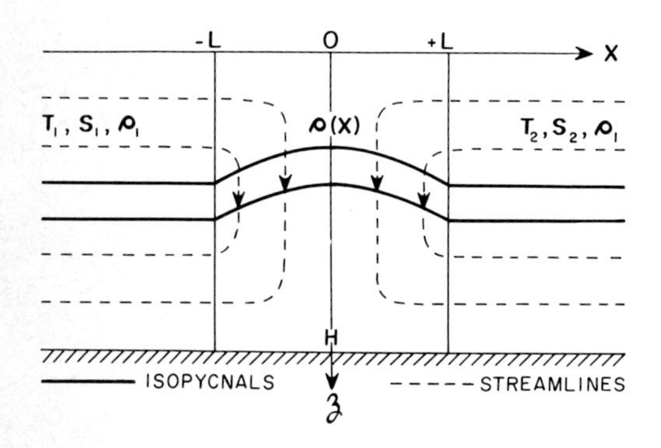

Fig. 3. Schematic diagram of a vertical frontal zone of width 2L, looking in the along-front direction.

analytical steady state cabbeling model for a narrow, vertical frontal zone of width 2L separating two water masses of equal density ρ. The cross-frontal length scale is assumed much less than R_{bc}, the Rossby radius of deformation; hence, Coriolis effects are neglected. The water is of depth H. Divergence is absent in the along-frontal direction. The advective terms in the equations of motion are neglected as velocities are expected to be very small.

Applying the Boussinesq approximation the equations of continuity and motion are

$$\frac{\partial u}{\partial x} + \frac{\partial w}{\partial z} = 0 \tag{1}$$

$$0 = -\frac{1}{\rho_1}\frac{\partial p}{\partial x} + N_x \frac{\partial^2 u}{\partial x^2} + N_z \frac{\partial^2 u}{\partial z^2} \tag{2}$$

$$0 = -\frac{1}{\rho_1}\frac{\partial p}{\partial z} + N_x \frac{\partial^2 w}{\partial z^2} + N_z \frac{\partial^2 w}{\partial z^2} + \frac{\rho - \rho_1}{\rho_1} g \tag{3}$$

where p is pressure, N_x and N_z are the horizontal and vertical eddy viscosities respectively, $\rho(x)$ is the density in the frontal zone, and ρ_1 is the constant far field density.

The equations of conservation of heat and salt are:

$$0 = K_x \partial^2 T/\partial x^2 + K_z \partial^2 T/\partial z^2 \tag{4}$$

$$0 = K_x \partial^2 S/\partial x^2 + K_z \partial^2 S/\partial z^2 \tag{5}$$

respectively, where K_x and K_z are the horizontal and vertical eddy diffusivities

The boundary conditions are:

$$S = S_1, \quad T = T_1 \quad \text{at} \quad x = -L$$

$$S = S_2, \quad T = T_2 \quad \text{at} \quad x = L \tag{6}$$

$\frac{\partial u}{\partial z} = 0$ at $z = 0, H$ no shear at surface and bottom

$w = 0$ at $z = 0, H$ no flux through surface and bottom

Equations (4) and (5) imply that heat and salt are mixed in the same way. Since T and S are independent of z, the solutions to equations (4), (5) and (6) are simply

$$\frac{T-T_1}{T_2-T_1} \equiv \frac{S-S_1}{S_2-S_1} \equiv \theta \quad \text{say}$$

$$= \tfrac{1}{2}(1 + x/L) \tag{7}$$

i.e., both temperature and salinity vary linearly across the front.

We now assume that the nonlinear density function across the frontal zone follows

$$\rho(T,S) = \rho_1(1 + k \sin \pi\theta),\ k > 0.$$

The constant $k \sim 10^{-6}$ is the amplitude of the density anomaly.

Using (7), we obtain

$$\rho(T,S) = \rho_1\{1 + k \sin \pi\ (1 + x/L)/2\},$$

so that the buoyancy acceleration is

$$\frac{\rho-\rho_1}{\rho_1}\, g = kg \sin \pi\ (1 + x/L)/2 \qquad (8)$$

Introducing the stream function ψ as $u = -\partial\psi/\partial z$, $w = \partial\psi/\partial x$ where u and w are the horizontal and vertical velocities, the final solution is (details are given in Bowman and Okubo, in press)

$$\psi' = \sigma\ \frac{\nu}{\nu-1}\ \frac{1}{(e^{\alpha}-e^{-\alpha})}\ \{(1-e^{-\alpha})\ e^{\alpha z'} + (e^{\alpha}-1)\ e^{-\alpha z'}\}$$

$$-\ \frac{1}{-1}\ \frac{1}{(e^{\beta}-e^{-\beta})}\ \{(1-e^{-\beta})\ e^{\beta z'} + (e^{\beta}-1)\ e^{-\beta z'}\}$$

$$-\ 1\ \cos \frac{\pi}{2}\ (1 + x')$$

where $\sigma = (\frac{2}{\pi})^4 \frac{\nu_z L^4}{\nu_x H^4}$, $\nu = \nu_x/\nu_z$

$$x = Lx' \quad , \quad z = Hz'$$

$$\psi = \frac{kg\pi}{2L}\ \frac{H^4\psi'}{\nu_z}$$

$$\alpha = \pi H/2L \quad , \quad \beta = \pi H\ \nu^{\frac{1}{2}}/2L$$

Setting

$H = 10^4$ cm, $L = 10^5$ cm, $\nu_z = 10^2$ cm^2 sec^{-1}, $\nu_H = 10^6$ cm^2 sec^{-1}, $k = 10^{-6}$, $g = 10^3$ cm sec^{-2}, then

$$\sigma = (2/\pi)^4 \quad , \quad \nu = 10^4$$

$$\partial\psi/\partial x = 10\ \partial\psi'/\partial x'$$

$$\partial\psi/\partial z = 10^2\ \ \partial\psi'/\partial z' \quad ,$$

we find that $w_{max} \sim .1$ mm sec^{-1} (~10 m day^{-1}) and $u_{max} \sim 1$ mm sec^{-1} (~100 m day^{-1}).

Streamlines are shown in Fig. 4. The associated velocities are too small to be measured directly, but in view of the persistent nature of the circulation, cabbeling may be a significant factor in maintaining a sharp interface. No inverted plume develops in our model, as contrasted to Foster's treatment, since we have neglected the advective terms in the equations of motion. Stated another way, horizontal diffusion mixes away the descending heavy water as fast as it forms and no plume develops. A linear model incorporating rotation is under development and will be submitted for publication elsewhere.

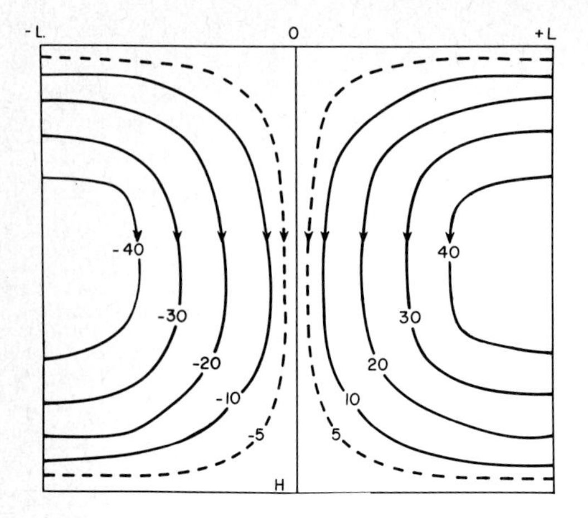

Fig. 4. Streamlines for the steady state cabbeling circulation in the frontal zone.

Conclusions

Further studies are needed to investigate the spatial and temporal variability of frontal circulation and the interacting water masses. It is worth repeating a list of suggestions in Stommel and Federov's (1967) paper, for the use of a servo controlled neutral float, locked into an intrusive feature as a tool to study the details of interleaving and cross-frontal mixing. "Among the things which we would like to do with such a float are the following:

(a) Photograph small dye streaks in homogeneous laminae, and in the gravitationally stable interfacial surfaces bounding them for periods of about an hour to obtain visual evidence of the nature of the regime.

(b) Photograph the relative (to the float) motion of small sinking floats to get the microprofiles (micro-spirals?) of velocity as well as temperature and salinity; and thus obtain a measure of local Richardson number.

(c) Repeat Pochapsky's 1961 experiments of dispersion of floats with more precise control of depth of floats.

(d) Measure frequency of interfacial waves from a float locked to an adjacent homogeneous lamina.

(e) Determine whether there is some correlation of micro-structural detail (for example interfacial shear) with the phase of large internal waves.

(f) Watch for tendency for unidirectional mixing across interfaces.

(g) Determine characteristic "lifetime" of laminae; for example, if laminae spread out they may become progressively thinner with time.

(h) Is there a "smallest" size of laminae?

(i) Look for very fine scale "salt convection" or other evidence of two-diffusivity convection with microconductivity cell mounted on arm attached to the float."

References

Bowman, M. J. and A. Okubo. Linear cabbeling model without rotation. To be submitted to J. Geophys. Res.

Foster, T. D. 1972. An analysis of the cabbeling instability in sea water. J. Phys. Oceanogr., 2, 294-301.

Horne, E. P. W. Interleaving at the subsurface front in the Slope Water off Nova Scotia. To be published in J. Geophys. Res.

Huppert, H. E. 1971. On the stability of a series of double diffusive layers. Deep-Sea Res., 18, 1005-1021.

Joyce, T. M. 1977. A note on the lateral mixing of water masses. J. Phys. Oceanogr. 4 (7) 626-629.

Linden, P. F. 1973. On the structure of salt fingers. Deep-Sea Res., 20, 325.340.

Oakey, N. S., J. A. Elliott. Vertical temperature gradient structure across the Gulf Stream. J. Geophys. Res., 82 (9) 1369-1380.

Osborn, T. R. and C. S. Cox 1972. Oceanic fine structure. Geophys. Fluid Dyn., 3, 321-345.

Pochapsky, T. E. 1961. Some measurements with instrumental neutral floats. Deep-Sea Res., 8, 269-275.

Stommel, H. and K. N. Federov. 1967. Small scale structure in temperature near Timor and Mindanao. Tellus, 19, 306-325.

Turner, J. S. 1965. The coupled turbulent transports of salt and heat across a sharp density interface. Int. J. Heat and Mass Transfer, 8, 759-769.

Turner, J. S. 1967. Salt fingers across a density interface. Deep-Sea Res., 14, 599-611.

Voorhis, A. D., D. C. Webb and R. C. Millard. 1976. Current structure and mixing in the shelf/slope water front south of New England. J. Geophys. Res., 81 (21) 3695-3708.

Witte, E. 1902. Zur Theorie der Stromkabbelungen. Gaea, Koln, 484-487.

APPENDIX A

Organizations Supporting the Workshop on the Roles of Oceanic Fronts in Coastal Processes

United States Environmental Protection Agency
Research and Development Office, Region II
26 Federal Plaza
New York, New York 10007

United States Energy Research and Development Administration
Marine Sciences Program
Environmental Programs
Division of Biomedical and Environmental Research
Washington, D.C. 20545

United States Coast Guard
Oceanographic Unit
Building 159-E
Navy Yard Annex
Washington, D.C. 20590

New York State Sea Grant Institute
State University of New York
99 Washington Avenue
Albany, New York 12246

Marine Sciences Research Center
State University of New York
Stony Brook, New York 11794

Office of Naval Research
Geography Programs
Code 462
Arlington, VA 22217

Stony Brook Foundation
P.O. Box 666
Stony Brook, NY 11790

APPENDIX B

Participants in the Workshop

Dr. William C. Boicourt
Chesapeake Bay Institute
The Johns Hopkins University
Baltimore, Maryland 21218

Dr. Malcolm J. Bowman
Marine Sciences Research Center
State University of New York
Stony Brook, New York 11794

Dr. Wayne E. Esaias
Marine Sciences Research Center
State University of New York
Stony Brook, New York 11794

Dr. Charles N. Flagg
E G & G, Environmental Equipment Division
Waltham, Massachusetts 02172

Dr. Robert D. Fournier
Institute of Oceanography
Dalhousie University
Halifax, Nova Scotia
CANADA B3H 4J1

Mr. Edward P. W. Horne
Institute of Oceanography
Dalhousie University
Halifax, Nova Scotia
CANADA B3H 4J1

Dr. Richard L. Iverson
Department of Oceanography
Florida State University
Tallahassee, Florida 32306

Dr. Christopher N. K. Mooers
Marine Studies Center
University of Delaware
P.O. Box 286
Lewes, Delaware 19958

Dr. Akira Okubo
Marine Sciences Research Center
State University of New York
Stony Brook, New York 11794

Mr. Bertram H. Olsson,
Defence Scientific Establishment
H. M. N. Z. Dockyard
Devonport, Auckland 9
NEW ZEALAND

Dr. Robin D. Pingree
Marine Biological Association
Citadal Hill
Plymouth, PL1 2PB
UNITED KINGDOM

Dr. John H. Simpson
Department of Physical Oceanography
University College of North Wales
Marine Science Laboratories
Menai Bridge
Anglesey, LL59 5EY
UNITED KINGDOM

Sponsor Representatives

Mr. Arnold Freiberger
U.S. Environmental Protection Agency
Research and Development Office, Region II
26 Federal Plaza
New York, New York 10007

Dr. Kenneth Mooney
U.S. Coast Guard, Oceanographic Unit
Building 159-E
Navy Yard Annex
Washington, D.C. 20590

Dr. J. R. Schubel
Marine Sciences Research Center
State University of New York
Stony Brook, New York 11794

B. Le Méhauté

An Introduction to Hydrodynamics and Water Waves

Springer Study Edition

1976. 231 figures, 12 tables. VIII, 323 pages
ISBN 3-540-07232-2

Contents: Establishing the Basic Equations that Govern Flow Motion. – Some Mathematical Treatments of the Basic Equations. – Water Wave Theories. – Appendices: Wave Motion as a Random Process. Similitude and Scale Model Technology.

"...The declared aim of the book is to provide a text based on a set of lecture notes for engineering students, that introduces the mathematical aspects of fluid mechanics with explanations of physical meaning to help practising engineers to read mathematical texts and keep pace with new developments reported in scientific journals... The text is divided into three main parts concerned, respectively, with the derivation of the basic equations of fluid motion, mathematical treatment of the equations, and water wave theories... Examples of clarity include well illustrated accounts of the distinctions between streamlines, stream tubes, streak lines and particle paths and of the significance of dilatation, shear and rotation in interpretating the kinematic equations. I specially commend the many well organised tabular summaries of key details and formulae which students will surely find very helpful and which offer convenient working references for practising engineers..."

N. Hogben in: Nature

Springer-Verlag
Berlin
Heidelberg
New York

Journal of Geophysics Zeitschrift für Geophysik

Edited for the Deutsche Geophysikalische Gesellschaft by W. Dieminger, J. Untiedt

Journal of Geophysics publishes articles predominantly in English from the entire field of geophysics and space research, including original essays, short reports, letters to the editor, book discussions, and review articles of current interest, on the invitation of the German Geophysical Association.
The following fields of geophysics have been treated in recent volumes: applied geophysics, geomagnetism, gravity, hydrology, physics of the solid earth, seismology, physics of the upper atmosphere including the magnetosphere, space physics and volcanology.

Fields of Interest: Geophysics, Seismology, Geomagnetism, Aeronomy, Extraterrestrical Physics, Space Research, Meteorology, Oceanography, Applied Geophysics, Theoretical Geophysics, Tectonics, Geochemistry, Petrology.

Subscription information and sample copies upon request.

Springer-Verlag
Berlin
Heidelberg
New York